PLANT ENGINEERING

Edited by **Snježana Jurić**

Plant Engineering
http://dx.doi.org/10.5772/65225
Edited by Snježana Jurić

Contributors

Nobuyuki Yoshikawa, Ichiro Kasajima, Norioko Yamagishi, Yekti Asih Purwestri, Febri Susanto, Hiroyuki Tsuji, Snježana Jurić, Redmond Ramin Shamshiri, Muhammad Razif Mahadi, AbdJamil Zakaria, Wan Ishak Wan Ismail, Desa Ahmad, Hasfalina Che Man, Kelly Thorp, Ramazan Beyaz, Mona Nagargade, Vishal Tyagi, Manoj Kumar Singh, Lev Elkonin, Valery Panin, Julia Italyanskaya, Dirk Balmer, Andrea Balmer, Emanuele De Paoli, Azeddine Si-Ammour, Brigitte Mauch-Mani, Ewa Mazur, Jiri Friml

Published by InTech
Janeza Trdine 9, 51000 Rijeka, Croatia

Publishing Process Manager Iva Simcic
Technical Editor SPi Global
Cover InTech Design team

First published November, 2017
Printed in Croatia
Legal deposit, Croatia: National and University Library in Zagreb

Additional hard copies can be obtained from orders@intechopen.com

Plant Engineering, Edited by Snježana Jurić
p. cm.
Print ISBN 978-953-51-3607-1
Online ISBN 978-953-51-3608-8

For EU product safety concerns:
IN TECH d.o.o., Prolaz Marije Krucifikse Kozulić 3, 51000 Rijeka, Croatia,
info@intechopen.com or visit our website at intechopen.com.

Meet the editor

Dr. Snježana Jurić currently works as a project advisor at the Ruđer Bošković Institute (RBI) in Zagreb, Croatia. Her scientific interests are strongly shaped around plant science; she is curious to discover how unknown proteins and protein interactions fit into the great fundamental processes, such as photosynthesis and flowering. Dr. Jurić completed her PhD thesis in Zagreb in 2010 and her postdoctoral study in Seoul under the mentorship of Prof. Ji Hoon Ahn, one of the world's leading researchers in the genetics of flowering, in 2014. She published several peer-reviewed articles; one chapter in the book *Electron Transfer Routes in Oxygenic Photosynthesis: Regulatory Mechanisms and New Perspectives*, published by InTech; and two chapters in the book *Methods in Molecular Biology*, published by RBI. Dr. Jurić is also interested in the project-oriented funding and development of leadership skills in science; she qualified as a professional specialist of project management in 2014. When not exploring the fascinating world of plants, she is reading science fiction literature and writing theatre play scripts.

Contents

Preface

This book is intended for those readers who are interested in recent discoveries in the field of basic plant science that have the potential to contribute to solving some of the greatest food issues today. The UN predicted that around 10 billion people will live on Earth till 2050. In order to feed the world, today's 2.1 billion tonnes of cereals will have to increase to at least 3 billion tonnes, according to FAO. Two major concepts are efficacy and sustainability. The modern agriculture has to explore more efficient strategies to grow additional food on a less available farm grounds due to the increased urbanization, poor quality of soil due to the drought and flooding events, political and economic fluctuations and constant decrease in the available number of rural labour forces. Simultaneously, it should embrace the new strategies to decrease the yield losses due to the diseasescaused by fungi, insects, nematodes, bacteria and viruses. There is a strong resistance against the GMO crops present in some parts of the world, mostly due to the mistrust in the high-profile companies that are producing and distributing the seeds, scepticism in scientific work, misunderstanding of the techniques utilized and poor or incomplete informing.From a basic plant biologist point of view, GMO is the logical step-up in the methodology from the classical plant breeding techniques, destined to develop even further. However, once the knowledge leaves the lab and becomes a know-how in the hands of biotech companies, we, as the world community, should demand stronger rules to be followed, constantly educate the younger generations and strive to develop even better and safer solutions. Although it is almost impossible to start a basic research today without any of the modern techniques involved, such as genetic engineering, in this book, there are a couple of chapters that describe traditional techniques revitalized. As scientists, we should strive not to be exclusive but to embrace and think about different strategies combined. Moreover, we should understand that none of the world-class problems will be solved overnight and that our best strategy is to plan and start basic plant science projects now to gather the knowledge tomorrow.

Dr. Snježana Jurić
Division of Molecular Biology
Ruđer Bošković Institute
Zagreb, Croatia

Section 1

Introductory Chapter

Chapter 1

Introductory Chapter: Plant Engineering for the Future Ahead

Snježana Jurić

Additional information is available at the end of the chapter

http://dx.doi.org/10.5772/intechopen.70489

1. Time to choose the right path

Earth today is experiencing visible climate changes, the one battle for which the humans are still not prepared enough to win. Existing economic, social, and cultural discrepancies between developing and developed countries are a constant obstacle for the "way forward," and the decisions about environment, food, and education brought today will influence our children tomorrow. This book is intended for those who still believe that our greatest strength is actually our curiosity. Basic research scientists worldwide are constantly struggling to obtain enough, mostly public, funds to provide sound foundations for innovations in all aspects of life; however, the pressure to publish first occasionally neglects the fact that every creative idea needs a considerable amount of time to be realized, often with negative results. Although the criticism in science is always welcomed, whether it comes from the fellow scientists or from the public voice, due to the lack of understanding, the complicated problems we are trying to solve become even more difficult. For example, if tomorrow the bacterium *Xylella fastidiosa*, outlined by Alyson Abbott in Nature (8 June 2017) as a huge problem for the olive groves in Southern Italy, spread to the rest of Europe, the only measure would be to uproot the contaminated plants. Although the infection could have been stopped/decelerated years ago, environmentalists challenged scientific findings and fight against contaminated olive plants uprooting. Italy has already allocated more than 5 million euros for the containment measures and will be forced to increase the funding in the future. We should ask ourselves, "Could we rely on science to find a more creative solution?"

2. Brief introduction to the chapters

The first chapter in the book is dedicated exactly to the problem of plant defence mechanisms against pathogens. According to the Food and Agriculture Organization of the United

Nations (FAO), each year between 20 and 40% world crop yields are lost due to the pathogens. One of the innovative tactics would be to explore the impact of the small RNA molecules on the plant-microbe interactions. Balmer et al. traveled beyond the well-documented antibacterial responses and concentrated on a hemibiotrophic fungus *Colletotrichum graminicola* that is detrimental for one of the most important world cereals, maize. In 1970, severe anthracnose epidemics eradicated the production of sweet corn in Indiana, USA. Disease was reported almost worldwide, with the yield loss ranging from 0 to 40%. Research on plant-microbe communication might help us to combat some of the worst pathogen-related crop diseases nowadays. Modulation of gene expression could be achieved by the usage of otherwise damaging gamma-rays, also. Beyaz and Yildiz exploited the power of gamma-irradiation to create new plant mutants with elevated amounts of antioxidant enzymes and proline. Although not novel, this technique is slowly showing its value as many countries express negative attitudes toward genetically modified foods (GMOs). One of the appealing strategies to increase the plant yield would be to obtain the control over the timing of flowering. FAO estimates that around 800 million people are currently undernourished. The food distribution and security throughout the globe is uneven and largely at risk due to the unpredicted weather events, extreme climate changes, diseases caused by pathogenic microbes, as well as abiotic diseases, and political perturbations. Flowering, as one of the fundamental processes in plants, has largely been investigated at the genetic level in the model plant Arabidopsis. Nevertheless, the number of reports on economically important plant species is steadily increasing, giving us the tool to artificially modulate the timing of transition from vegetative to reproductive plant stage. In that sense, Purwestri et al. identified the proteins that are able to interact with the protein responsible for the transition from vegetative to reproductive phase, the Flowering Locus T (FT), in rice. According to the International Rice Research Institute (IRRI), more than half of the 7 billion people on the planet consume rice as the staple food. Kasajima et al. demonstrated an interesting approach to accelerate the flowering time by introducing the Apple latent spherical virus vector in a number of plant species; again, an interesting approach to circumvent the bad reputation toward GMOs. Not only the food availability but also its usability could be improved. In the third chapter, genetic transformation of the sorghum plant led Elkonin et al. to improve the nutritional value of this highly important crop. Mazur and Friml investigated in great detail vascular tissue development and regeneration in Arabidopsis. This knowledge would be beneficial in the future tree investigations. Finally, the fourth chapter deals with the constant efforts of the scientific community to predict and battle the unfavorable living conditions on Earth. Nagargade et al. discuss about the strategies to develop the climate-resilient agriculture. One of the approaches for increased yield would be to "teach" the greenhouses to constantly change the parameters in order to achieve the optimal growing conditions for the given plant species, as described by Shamshiri et al.

All the strategies described in this book represent a small, but valuable contribution of the plant scientific community to decrease the yield losses, improve the food nutritional values, and make it safer for consumption. Only through the joint effort and the constant collaboration between the scientists, farmers, and distributers, we could gain the consumers' trust and implement the smart managing over the Earth resources.

http://dx.doi.org/10.5772/intechopen.70489

Author details

Snježana Jurić

Address all correspondence to: sjuric@irb.hr

Ruđer Bošković Institute, Zagreb, Croatia

Section 2

Remodeling the Plant Defence Respons: New Approaches

Chapter 2

Signs of Silence: Small RNAs and Antifungal Responses in *Arabidopsis thaliana* and *Zea mays*

Andrea Balmer, Emanuele De Paoli,
Azeddine Si-Ammour, Brigitte Mauch-Mani and
Dirk Balmer

Additional information is available at the end of the chapter

http://dx.doi.org/10.5772/intechopen.69795

Abstract

Plant small RNAs (sRNAs) are pivotal regulators of gene expression, which are crucial in maintaining genome integrity and flexibility during development, abiotic and biotic stress responses. Current evidence suggests that sRNAs might be inherent to the sophisticated plant innate immune system battling bacteria. However, the role of sRNAs during antifungal plant defences is less clear. Therefore, this chapter investigates the sRNA-mediated plant antifungal responses against the hemibiotrophic fungi *Colletotrichum higginsianum* and *Colletotrichum graminicola* in their respective compatible hosts *Arabidopsis thaliana* and Zea mays. A phenotypic and metabolomic analysis of *A. thaliana* sRNA mutants in response to *C. higginsianum* infection was performed, showing a hormonal and metabolic imbalance during fungal infection in these plants. To find whether fungal-induced sRNA could directly regulate defence genes in an agricultural important plant model, the expression of maize miRNAs in response to *C. graminicola* leaf and root infections was investigated. The results revealed the tissue-specific local and systemic adaptation of the miRNA transcriptome, where only a few miRNAs were targeting defence pathways. The general picture presented here points towards a role of sRNAs as fine-tuners of genetic and metabolomic defence response layers. This chapter also further discusses the potential of utilizing sRNA-based fungal control strategies.

Keywords: small RNA, antifungal plant defence, metabolomics, deep sequencing

1. Introduction

Small RNAs (sRNA) are small noncoding RNA segments of 19–30 nucleotides in length [1]. They mediate gene silencing, a gene regulation mechanism acting on a transcriptional

(transcriptional gene silencing (TGS)) and post-transcriptional level (post-transcriptional gene silencing (PTGS)). In general, sRNA molecules originate from the transcription of endogenous microRNA (miRNA genes), other genomic sRNA loci, aberrant RNA produced by transposons as well as invasive viral RNA [2]. Plants carry two main classes of sRNAs grouped according to their size, function and biogenesis, namely microRNAs (miRNA) and short-interfering RNAs (siRNA) [3]. Such sRNAs are generated through various mechanisms; within the miRNA biogenesis pathway, miRNA precursors derived from MIR genes are processed in the nucleus by Dicer-like protein 1 (DCL1) and exportin-like protein (HYL1) into mature miRNA duplexes of 20–22 nucleotides in length. Mature miRNAs are then methylated at the 3′ terminus by HEN1 (small RNA methyltransferase) and exported to the cytoplasm. One strand of the duplex is incorporated into an argonaute protein (AGO) protein to form an RNA-induced-silencing complex (RISC) [4]. The siRNAs, however, originate from long dsRNA that can be derived from transgenes, viruses, transposons and natural sense-antisense transcripts. Such long dsRNA is recognized and cleaved by a certain type of DCL proteins; thereby siRNA classes with different sizes are generated. Like miRNAs, siRNAs are loaded into an AGO protein-containing RISC that controls gene expression patterns through the degradation of mRNA or the repression of translation of fully/partly complementary sequences of mRNAs, as well through epigenetic changes via mediation of DNA and histone methylation [5, 6].

Gene silencing is not only important for the maintenance of genome integrity by silencing transposons or by degrading the viral RNA but also important during host immune responses of both plants and animals [7–9]. The recognition of pathogens by plants leads to the activation of a multi-layered immune system that comprises the establishment of a complex network of inducible defences including pathogen-associated molecular pattern (PAMP)-triggered immunity (PTI) and effector-triggered immunity (ETI) [10, 11]. The entire signalling process involves the regulation of defence gene expression, the release of plant hormones and/or the induction of secondary metabolites [12]. Over the past few years, plant sRNA pathways were recognized as important players during PTI and ETI [13, 14]. In Arabidopsis, bacteria-induced miRNAs were identified to orchestrate components of plant hormone signalling, including auxin, abscisic acid (ABA), jasmonic acid (JA) and salicylic acid (SA) [15, 16]. A canonical example of an miRNA regulating plant defence is miR393. It is up-regulated upon treatment with a bacterial PAMP, and negatively regulates auxin signalling and therefore contributes to SA-mediated PTI responses in Arabidopsis [17].

Although the important role of sRNAs in plant defence against viruses and bacteria is documented [8, 13], their function as components of the plants' defence response against fungi is less clear. Advances in genome-wide studies revealed a massive adaptation of host miRNA expression patterns after infection by fungal pathogens such as *Fusarium virguliforme* [18], *Erysiphe graminis* [19], *Verticillium dahliae* [20], *Cronartium quercuum* [21], as well as the oomycete *Phytophthora sojae* [22]. The alterations in sRNA expression profiles upon fungal attack suggest that gene silencing also contributes to antifungal defence; however, up to date there are no putative mechanisms deciphered. Besides orchestrating plant defence, sRNA could also act as direct antifungal molecules, as some plant miRNAs could share complementarity to fungal genes. This possibility has already been demonstrated by engineering transgenic plants expressing dsRNA targeting fungal genes and exhibiting enhanced resistance to different fungi. For instance, this mechanism named host-induced gene silencing (HIGS) was successfully applied for various plant-fungi pathosystems such as silencing of the *Blumeria graminis* effector Avra10 [23], or *CYP51* genes of *F. graminearum* [24].

In this study, we aim to elucidate the role of sRNAs in regulating susceptibility to *Colletotrichum* spp.; hence we congregated results from two compatible pathosystems: *C. higginsianum*, which infects plants from the Brassicaceae family such as *Arabidopsis thaliana* (**Figure 1**) and *C. graminicola*, which is a devastating pathogen of the industrially important crop *Zea mays* (**Figure 2**). Both ascomycetes use a multistage hemibiotrophic strategy to infect their host and also share close genetic similarities making them tractable models to compare fungal pathogenicity in both dicot and monocot models [25–28]. In Arabidopsis, *C. higginsianum* employs first a biotrophic stage limited to a confined array of first invaded cells, from where the fungus develops secondary hyphae to switch to necrotrophic growth into surrounding cells. *C. graminicola* extends the biotrophic lifespan into many host cells, persisting biotrophic at the margins, whereas the centre of infection becomes necrotrophic. *C. graminicola* is a major worldwide threat for corn cultures, as it affects all parts of the plants, either as leaf blight or as stalk rot [29]. Depending on specific corn hybrids and culture conditions, *C. graminicola* can result in up to 40% yield loss where endemic.

During the first step, a selection of sRNA mutants and two fully and intermediate fungal susceptible accessions of *A. thaliana* was examined in order to dissect possible defence defects caused by mutations in sRNA biogenesis pathways. Thus, we analysed the accumulation of phytohormones that are known to mediate Arabidopsis resistance against *C. higginsianum* [30] and secondary metabolites that function as direct defences [31]. We show that some Arabidopsis sRNA mutants display an altered susceptibility against *C. higginsianum*,

Figure 1. *Arabidopsis thaliana* leaves infected by *Colletotrichum higginsianum*, 6 days post inoculation.

Figure 2. *Zea mays* leaf (left) and root (right) infected with *Colletotrichum graminicola*.

together with a defective setup of chemical defences. Moreover, to better understand the role of sRNA during infection with *Colletotrichum* spp., we performed an miRNA expression profiling to obtain a deeper insight into adaptations of the sRNA transcriptome in different *C. graminicola*-infected maize tissues. The miRNA profiling demonstrated that the vast majority of altered miRNAs were targeting genes that are not directly linked to antifungal-defence pathways, suggesting that antifungal-defence responses are not regulated by specifically induced miRNAs.

This chapter provides a multi-omics analysis of sRNA-mediated antifungal plant reactions on a phenotypic, metabolomic as well as transcriptomic point of view. Altogether, our data propose a rather indirect defensive role of sRNAs in calibrating metabolomic and transcriptomic balances during antifungal responses against *Colletotrichum* spp. Future putative applications of sRNA-based fungal control strategies will be commented.

2. Materials and methods

2.1. Plant material and growth conditions

A. thaliana genotypes (*hen1-1, hyl1-2, rdr6-15*, Col-0 and Ler-0) were germinated in soil maintained at 21°C day/20°C night, with 9 h of light (120 μE $m^{-2}s^{1}$) and 60% of relative humidity. Selected *A. thaliana* accession Ler-0 was described to be susceptible to *C. higginsianum* infection, while Col-0 showed intermediate resistance [31]. Ler-0 is the wild-type genetic background of *hen1-1* mutants; all other mutants have a Col-0 genetic background. One week after germination, seedlings were individually transferred to 33-mL Jiffy pellets and kept in the same conditions until the infections. *Z. mays* (variety Jubilee, West Coast Seeds, www.westcoastseeds.com) was cultured in a soil-free plant growth system as described by Ref. [32].

2.2. Pathogen and pest cultivation and inoculation

C. higginsianum IMI34 349061-GFP [26] was cultured on potato dextrose agar (PDA) in a growth chamber under permanent light at 25°C. For infections, a fungal spore suspension of 10^6 spores mL^{-1} was prepared from 2-week-old cultures. Four- to five-week-old *A. thaliana* plants were drop-inoculated with 5 μL of the spore suspension. The plants were then incubated in darkness for 16 h at 25°C and 100% relative humidity. Post incubation, the growth condition of the plants was changed to long day (16 h/8h day/night cycle at 25°C). Control plants were treated only with sterile water. *C. graminicola* M1.001 was cultivated on PDA under permanent light at 25°C; infection assays were performed on 12-day-old maize plants as previously described [32].

2.3. Quantification of fungal growth

In planta fungal growth of *C. higginsianum* was measured every 24 h post infection for 4 days. The infection sites of the green fluorescent protein-expressing fungal strain were illuminated using a Nikon C-SHG1 UV lamp. Images were captured using a Nikon DS-L1 camera and the pictures were further analysed with the help of ImageJ (http://rsbweb.nih.gov/ij/) and Adobe Photoshop CS3 (http://labs.adobe.com). The area of fungal growth was measured in pixels and converted to mm^2.

2.4. Hormone quantification

For hormone analysis, salicylic acid, jasmonic acid and abscisic acid were quantified simultaneously from leaf material using UHPLC-MS/MS as described [32]. Hormone measurements were performed 4 days post *C. higginsianum* infection. To analyse each Arabidopsis accession, three independent biological replicates per sample were generated, each replicate a pool of five plants.

2.5. Metabolomic profiling

For metabolomic analysis, 4-week-old Arabidopsis plants were infected with *C. higginsianum*. Metabolites were isolated and analysed 4 dpi as described [32]. Six technical replicates for each treatment were analysed, and each replicate consisted of a pool of four plants.

2.6. Gene expression analysis

Confirmation of down-regulation of maize genes putatively targeted by miRNAs was conducted as described [32], using ZmGAPc as normalizing gene. Primer sequences are as follows: ZmATPS_fw: tcgtattaatgctggtgcaaac, ZmATPS_rev: ctctgtggggtggctcat; ZmSAT_fw: ttataaaaaccctgttcttctgctc, ZmSAT_rev: aggacaccttcctcaagaacc; ZmGAPc_fw: gcatcaggaaccctgaggaa, ZmGAPc_rev: catgggtgcatctttgcttg.

2.7. Deep sequencing and Northern blotting of maize sRNAs

For sRNA library preparation, six biological replicates were pooled and total RNA was isolated using Trizol (Invitrogen, www.invitrogen.com); 10 µg of total RNA was further processed using an Illumina-Solexa deep-sequencing approach at FASTERIS (http://www.fasteris.com). The expression of selected miRNAs was further analysed using sRNA Northern blotting techniques as described [33].

2.8. Identification and quantification of conserved miRNAs

To identify conserved maize miRNAs, sequences of 4677 mature plant miRNAs were downloaded from miRBase (release 18.0, November 2011). Identical miRNA sequences identified in different species or duplicated loci in a genome were collapsed, resulting in a non-redundant list consisting of 2228 unique miRNAs. Sequences belonging to the same miRNA family were further analysed by multiple alignment using ClustalW (www.clustal.org) and classified in subgroups to distinguish bona fide mature miRNAs from misannotated miRNA* forms or sequences generated from different regions of the same precursor. This non-redundant library was then applied to screen the small RNA libraries. All the small RNA reads in the range of 20–24 nt in size, and which are represented and represented by at least two reads in a library were aligned to the 1772 unique miRNAs derived from miRBase. For the screening, a maximum of three mismatches was allowed and up to 2 nt overhanging nucleotides at the 5′ and/or 3′ end. Alignments were performed using SeqMap [34]. The output was filtered and reformatted with custom PERL scripts, classifying the identified miRNAs according to miRBase.

2.9. Target prediction of maize miRNAs

Putative targets of maize miRNAs were identified using the psRNATarget web server (http://bioinfo3.noble.org/miRU2/) against *Z. mays* DFCI Gene index (version 19) and *Z. mays* PlantGDB genomic project. Default settings were applied.

2.10. Statistical analysis

Variances of quantified levels of metabolites and fungal growth for multiple groups were analysed by a one-way analysis of variance (ANOVA); a *P*-value of <0.05 was considered significant. The Mann–Whitney *U*-test was used to compare significant differences between two sample groups. All statistical analysis was performed using Sigma Plot 11.0 (http://www.sigmaplot.com).

3. Results

3.1. Arabidopsis sRNA mutants show different levels of susceptibility to *C. higginsianum*

To test if a functional silencing machinery is required for a proper antifungal-defence response, *A. thaliana* wild types Ler-0 and Col-0 showing lower and intermediate resistance, respectively, and sRNA pathway mutants were subjected to fungal infection assays to monitor the susceptibility to *C. higginsianum*. To cover important components of sRNA pathways, the loss of function mutants for the genes encoding *HYL1, HEN1*and *RDR6* was analysed. The sRNA pathway mutants were infected with *C. higginsianum*-GFP, and the disease progression was compared to the relative wild-type ecotype, for *hen1-1* namely *Landsberg erecta* (Ler-0), for all other mutants Columbia (Col-0). Fungal growth was monitored at 24, 48, 72 and 96 h postinfection (hpi) (**Figure 3**). These time points were chosen to cover all known infection stages of *C. higginsianum* during hemibiotrophic growth on leaves [25, 29]. The infection assays showed an altered susceptibility of mutants (**Figure 3**). For *hen1-1,* a significant higher susceptibility was only detected in late infection stages (96 hpi). Comparison of *hyl1-2* with Col-0 yielded statistically significant differences of fungal growth at all time points (**Figure 3(b)**). The RNA mutant was found to be more susceptible to *C. higginsianum* compared to the wild type. By contrast, *rdr6-15* was infected by *C. higginsianum* as efficiently as the wild type (**Figure 3(c)**). Altogether, a defective sRNA machinery seems to render plants more susceptible to fungal attack. However, mutations in *RDR6-15* did not alter the susceptibility against the *C. higginsianum*.

3.2. Arabidopsis sRNA mutants show an altered hormonal balance after *C. higginsianum* infection

Hormone signalling is a key process that regulates stress responses. To evaluate the implication of sRNA pathways in hormone-mediated plant defence against *C. higginsianum,* levels of salicylic acid, jasmonic acid and abscisic acid were quantified by HPLC-MS/MS. All selected mutants and wild-type accessions were analysed 4 days post *C. higginsianum* infection and hormone levels of both infected and mock were measured. In response to *C. higginsianum* attack, SA and JA were induced to different levels in all genotypes (**Figure 4**). Notably, SA and JA inductions were more pronounced in the mutants *hen1-1* and *hyl1-2* compared to their respective wild-type (Ler and Col-0) infected plants. For instance, in infected *hen1-1* plants, JA levels rose up to 589 ng/100 mg fresh weight, whereas in infected Ler plants, JA only reached 234 ng/100 mg fresh weight. However, *rdr6-15* did not appear to have significant differences of SA and JA levels compared to wild-type-infected plants (**Figure 4(a)(b)**). On the other hand, ABA levels were found to be induced during fungal infection in *hyl1-2* and *hen1-1* contrary to *rdr6-15* that show no significant changes in ABA quantity upon fungal infection (**Figure 4(c)**). These results suggest that the sRNA mutant *rdr6-15* is likely not implicated in the regulation of hormone levels during antifungal responses, whereas a functional HEN1 and HYL1 protein seems to be required to mount a full SA, JA and ABA response to fungal attack.

3.3. The metabolome of Arabidopsis sRNA mutants in responses to *C. higginsianum* infection

To compare the changes in the metabolomic profile of sRNA mutants and wild-type plants induced by *C. higginsianum* infection, an UHPLC-QTOF-based analysis of secondary metabolites was performed. The metabolomic fingerprinting provided a global view on the metabolic perturbations induced by *C. higginsianum* attack at 4 dpi. Comparison of the metabolome *hen1-1* and *hyl1-2* mutants by a principal component analysis (PCA) resulted in a clear separation of both control-treated and -infected mutants and their respective wild types (**Figure 5(a)** and **(b)**).

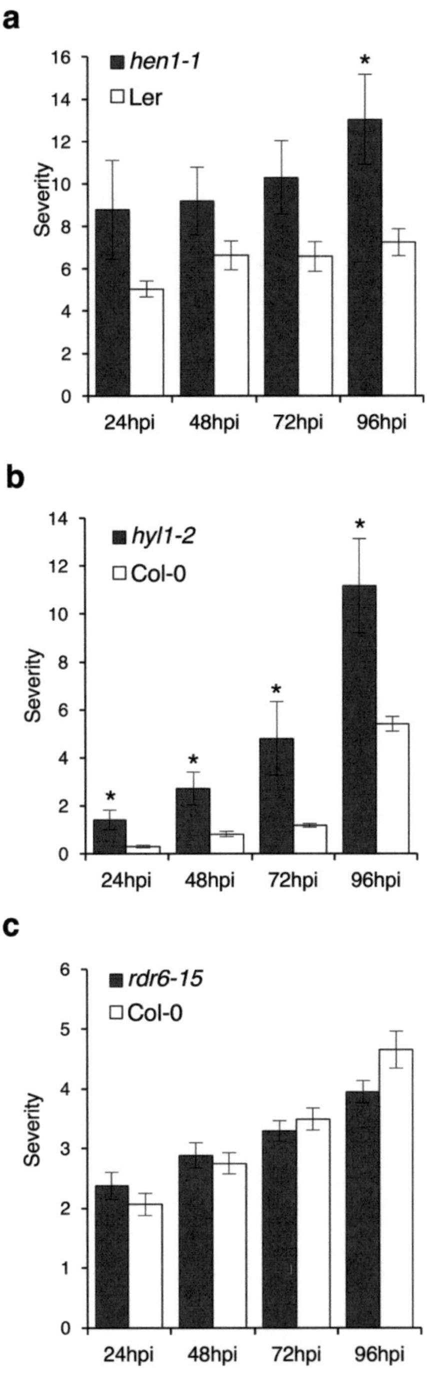

Figure 3. Disease severity of *C. higginsianum* in *A. thaliana* sRNA mutants and wild-type plants; (a). *hen1-1* mutant, (b). *hyl1-2* mutant, (c). *rdr6-15* mutant, compared to the respective wild-type background. Fungal growth was determined by quantifying the fluorescent area of *C. higginsianum*-GFP in mm^2 at different time points in all *A. thaliana* mutants compared to wild type. Severity was determined as percentage of leaf area affected. For statistical analysis, a one-way ANOVA was applied; asterisks indicate statistically significant differences ($P < 0.05$). Error bars indicate standard deviation (SD).

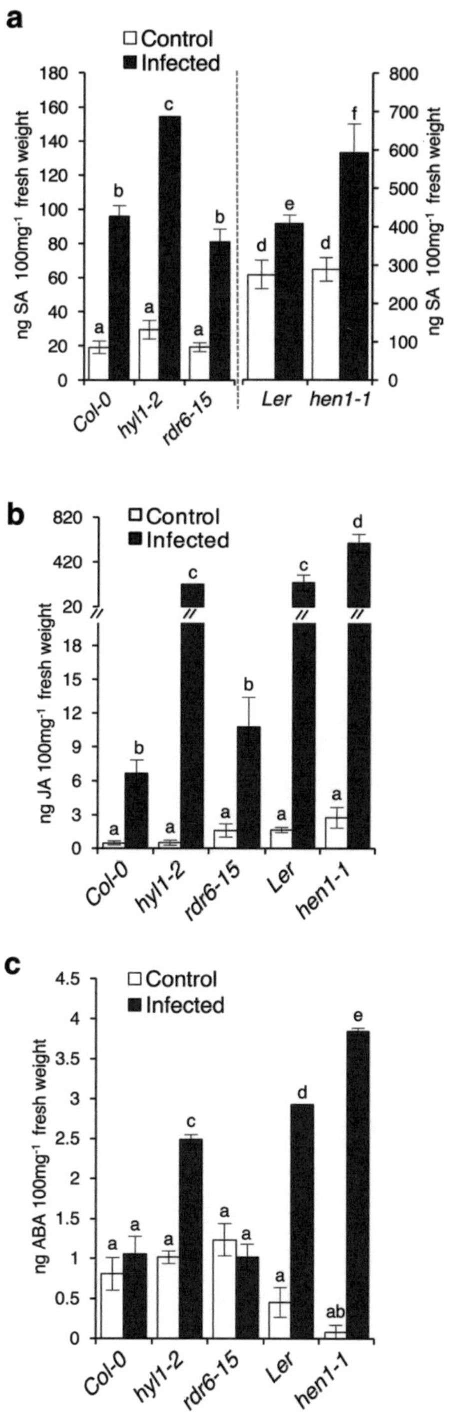

Figure 4. Quantification of phytohormones in *A. thaliana* sRNA mutants and wild-type plants after *C. higginsianum* infection. (a). salicylic acid (SA), (b). jasmonic acid (JA), and (c). abscisic acid (ABA) in *A. thaliana* sRNA mutants (*hen1-1, hyl1-2,* and *rdr6-15*) and wild-type plants (Col-0 and Ler) under two treatments: infected with *C. higginsianum* and control. Statistical significance was determined using one-way ANOVA. Letters indicate statistically significant differences ($P < 0.05$). Error bars indicate standard deviation (SD).

The PCA performed for *rdr6-15* grouped the mutants and wild type much closer (**Figure 5(c)**). The metabolomic fingerprinting allowed identifying groups of putative antifungal metabolites that were normally induced in the wild type, for which in turn the mutants showed an abnormal induction pattern. After PCA analysis, the compounds showing the greatest difference between wild type and mutants were selected for further identification (**Table 1**). Compounds were identified by exact mass, fragmentation spectrum and the retention time of the fragments using the online free databases Metlin, MassBank, Kegg and Aracyc and the in-house database from the chemical analytical service of the University of Neuchatel. The metabolomic analysis revealed a group of glucosinolates, flavonols, phenylpropanoids and the phytoalexine camalexine that were differentially induced in mutants and wild-type plants (**Table 1**). In response to *C. higginsianum*, *hen1-1* mutant showed lower fold induction of some glucosinolates like 7-methylthioheptyl glucosinolate, glucoerucin, glucoiberin, glucoiberverin and glucolesquerellin. Moreover, glucobrassicin was not induced after infection in *hen1-1* plants. Kaempferol 3-O-rhamnoside-7-O-rhamnoside (kaempferol 3-rha-7-rha) and kaempferol 3-O-rhamnoside-7-O-glucoside (kaempferol 3-rha-7-glu), flavonols which are well-described antifungal compounds [35], were down-regulated in *hen1-1* and Ler plants as well as the phenylpropanoids sinapoyl malate and 1-O-β-D-glucopyranosyl sinapate. The phytoalexin camalexin was the most induced compound after infection in *hen1-1* and Ler plants. Ler showed 84.4-fold induction of camalexin while infected *hen1-1* contained 10.7 more than mock-treated plants. The *hyl1-2* mutant exhibited lower fold induction in most of the glucosinolate levels compared to Col-0 (**Table 1**). Glucobrassicin, glucoiberin and glucoiberverin levels were higher in *hyl1-2* control and infected treatments than in wild type plants. Moreover, the induction of kaempferol 3-rha-7-rha and kaempferol 3-rha-7-glu was higher in Col-0 than *hyl1-2* plants. Levels of sinapoyl malate and 1-O-β-D-glucopyranosyl sinapate were also lower in *hyl1-2* control and infected plants compared to Col-0. Camalexin was 72.9-fold induced in Col-0 and 69.0-fold induced in *hyl1-2*. The *rdr6-15* mutant exhibited lower fold induction of all glucosinolates, flavonols and phenylpropanoids mentioned in **Table 1** compared to Col-0. The fold induction of camalexin was similar in *rdr6-15* mutant compared to Col-0 plants. In summary, sRNA mutant *hen1-1* exhibited lower levels of pathogen-induced camalexin, whereas the glucosinolates, flavonol and phenylpropanoid compounds were slightly less prominently induced in response to fungal infection in all the mutants compared to their respective wild-type plants.

3.4. *C. graminicola*-infected maize sets up a tissue-specific miRNA profile which is not directly linked to plant defence

Using annotated maize miRNAs (zma), known miRNAs were classified in the different maize sRNA libraries. In order to determine biostress-specific miRNAs and to quantify their expression level in the treated samples, the fold change expression was determined by calculating the relative difference of sequence reads in treated samples compared to the control libraries. Selected miRNAs showing a fold change of >2 are summarized in **Table 2**. Comparing biotrophic and necrotrophic fungal infection stages to mock, zma-miR479, zma-miR1318 and zma-miR1432 were found to be up-regulated; however, their fold induction was higher during the necrotrophic stage. Other miRNAs such as zma-miR393, zma-miR1120 and zma-miR2092 showed an altered expression level exclusively during the biotrophic stage. By contrast, the

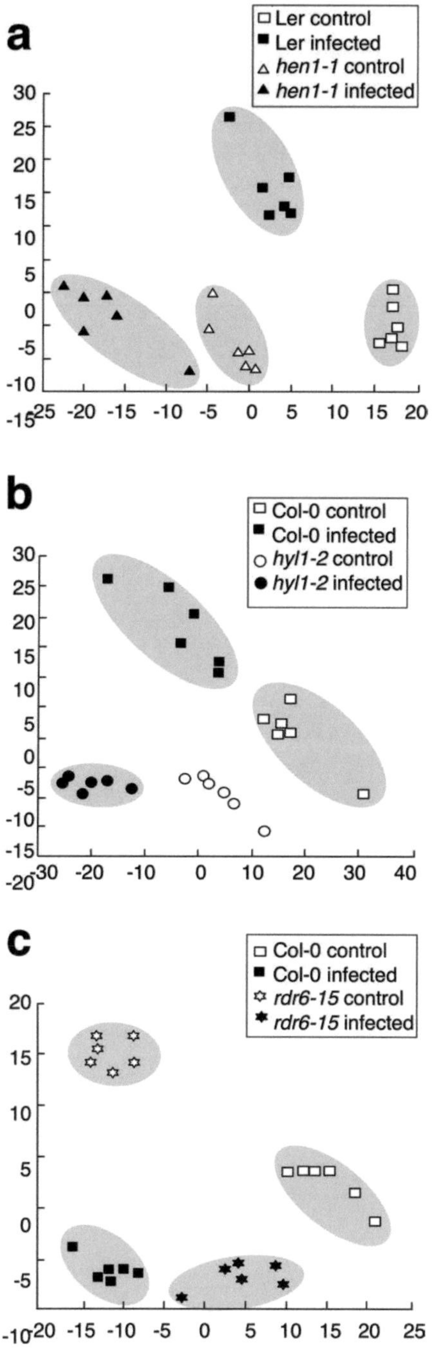

Figure 5. Metabolites distribution in sRNA mutants and wild-type plants upon *C. higginsianum* infection and control treatment. Principal component analysis (PCA) score plot of the metabolome. of the sRNA mutants *hen1-1* (a), *hyl1-2* (b), *rdr6-15* (c) and the wild-type Ler and Col-0 upon 4 dpi with *C. higginsianum* infection and control treatment. The PCA analyses were performed using Marvis Filter and Cluster packages, following a Kruskal-Wallis test ($P < 0.05$). Each data point represents one replicate of six independent biological replicates.

Compound	Mass	Fragments (M-H)-	*hen1-1* FI	*Ler FI*	*hyl1-2* FI	*rdr6-15* FI	Col-0 FI
Glucoberteroin	434.0612	96.9603, 95.9523	-	-	0.4	0.9	1.9
Glucobrassicin	447.0512	96.9601, 95.9523, 74.9914	0.8	1.7	2.0	0.6	2.4
Glucoerucin	420.0457	96.9628, 95.9551, 74.9943	1.0	1.6	0.2	0.8	1.7
Glucoiberin	422.0219	96.9619,95.9519, 74.9923	0.8	1.4	1.7	0.7	1.6
Glucoiberverin	406.0301	96.9619, 95.9494, 74.9920	1.1	2.1	1.0	0.7	1.4
Glucolesquerellin	448.0764	96.9590, 95.9513,74.9919	1.1	1.6	1.4	1.1	2.0
Gluconasturtiin	422.0578				0.8	0.8	2.1
Glucoraphanin	436.0406	372.0467, 178.0225	0.7	1.1	0.7	0.8	1.8
7-Methylthioheptyl glucosinolate	462.0958	95.9527, 74.9920	1.0	1.6	1.2	0.9	1.9
kaempferol 3-O-rhamnoside-7-O rhamnoside	578.1552	431.0942, 285.0399, 283.0236	0.6	1.0	1.4	0.8	1.7
kaempferol 3-rhamnoside-7-Glu	593.1534	447.0905, 285.0410, 283.0240	0.6	0.9	0.8	0.7	1.4
Sinapoyl malate	339.0745	223.0586, 164.0484, 149.0245	0.7	0.9	1.2	0.8	1.4
1-O-β-D-glucopyranosyl sinapate	385.1147	265.0794, 190.0267, 175.0030	0.8	1.0	0.9	0.5	1.2
Camalexin	199.0332		10.7	84.4	69.0	71.5	72.9

Fold induction of identified compounds from the metabolome of the sRNA mutants *hen1-1, hyl1-2, rdr6-15* and the wild-type Ler and Col-0 upon *C. higginsianum* infection (4 dpi).

Table 1. Fold induction of metabolites in sRNA mutants and controls upon *C. higginsianum* infection.

http://dx.doi.org/10.5772/intechopen.69795

expression of zma-miR168, zma-miR2916 and zma-miR5205 was altered only during the necrotrophic stage. Notably, zma-miR1432 and zma-miR2092 were also up-regulated in infected roots, suggesting that some miRNAs are regulated organ independently. Notably, infected roots showed also a distinct expression profile with zma-miR166, zma-miR169 and zma-miR395 that were down-regulated, whereas zma-miR909 and zma-miR2863 were up-regulated. A different situation was found in systemic leaves upon leaf infection. Compared to local infected tissues, less miRNAs showed an altered expression. For instance, zma-miR397, zma-miR916 and zma-miR5169 were up-regulated. In systemic leaves upon root infection, zma-miR1877 and zma-miR2592 were down- and up-regulated, respectively. Interestingly, zma-mi395 was down-regulated, and zma-miR479 showed elevated expression levels; zma-miR479 was also found to be up-regulated in local leaf infections, whereas the down-regulation of zma-miR395 was also observed in infected roots. In summary, although some miRNAs were commonly regulated in both locally infected leaves and roots, the miRNA transcriptome was specific for a given infection stage and in addition also organ-specific (**Table 2**). To confirm the deep-sequencing results, Northern blots of a selected miRNA were performed. Due to the relatively high expression level and the remarkable difference between control and treated samples, zma-miR395 was selected (**Figure 6**).

Library	miRNA	FI	Putative target genes
Inf L 24h	miR393	2.23	Calmodulin-binding protein MPCBP; cyclin-like F-box
	miR479	3	Unknown
	miR1120	−3	Unknown
	miR1432	2.3	Para-hydroxybenzoate-polyprenyltransferase (LOC100282174)
	miR2092	7	Unknown
Inf L 96h	miR168	2.7	Argonaute and Dicer protein; ubiquitin carboxyl-terminal hydrolase—*Oryza sativa*
	miR479	4	Unknown
	miR1432	18.3	Para-hydroxybenzoate-polyprenyltransferase (LOC100282174)
	miR2916	3.3	Quinone reductase 2—*Triticum monococcum; Zea mays* 18S ribosomal RNA gene
	miR5205	−3.25	Unknown
Inf R 96h	miR166	−6.5	MFS14 protein precursor; basic-leucine zipper (bZIP) transcription factor; lipid-binding

Library	miRNA	FI	Putative target genes
	miR169	−3.8	RAPB protein—*Oryza sativa*; allene oxide synthase—*Zea mays*
	miR395	−15.5	ATP sulphurylase (LOC541653), mRNA
	miR909	5	Inhibin, beta B subunit; vinculin; heavy metal transport/detoxification protein
	miR1432	4.5	Para-hydroxybenzoate-polyprenyltransferase (LOC100282174)
	miR2092	2.6	Unknown
	miR2863	3.5	Unknown
Inf L sys L	miR397	2.2	Laccase; multicopper oxidase;
	miR916	3.3	Zein protein-body ER membrane protein
	miR5169	2.2	Unknown
Inf R sys L	miR395	−2.7	ATP sulphurylase (LOC541653), mRNA
	miR479	3	Unknown
	miR1877	−3	Putative protein binding protein
	miR2592	3	Unknown

FI = fold induction compared to control libraries. Inf = infected, L = leaf, R = root, sys = systemic, zma = maize miRNAs.

Table 2. Maize miRNAs differently regulated upon *C. graminicola* infection.

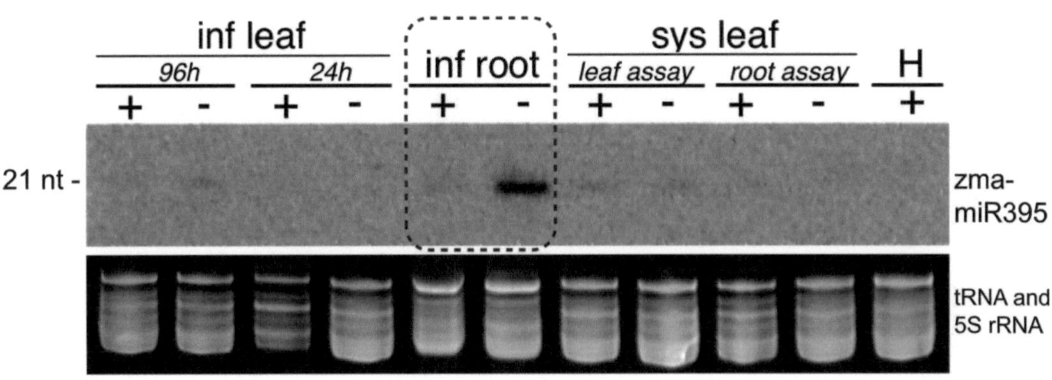

Figure 6. Northern blot analysis of miR395 expression. The signs + indicates *C. graminicola* infection, - control tissue. H= herbivore (*Spodoptera frugiperda*, non-fungus control). The tRNA and 5S rRNA are shown as a control for equal loading and were stained with ethidium bromide.

http://dx.doi.org/10.5772/intechopen.69795

As expected, zma-miR395 showed a reduced expression level upon fungal infections in roots. The signal intensity also corresponded to the sequence reads in the different libraries, with the highest number of reads (93) in control roots. To examine the putative role of zma-miR395 during root infections, the maize genome was analysed for putative target genes. Five known target genes were identified: two genes (dienelactone hydrolase and FMR1-interacting) exhibit two mismatch positions for zma-miR395. The other genes, ATP sulphurylase (APS) on chromosomes 1 and 5, and a sulphate anion transporter, perfectly matched to the zma-miR395 sequence. To confirm the genotype of a reduced expression level of zma-miR395 in infected maize roots, the gene expression of two zma-miR395 putative target genes (*ZmSAT* and *ZmATPS*) was analysed (**Figure 7**).

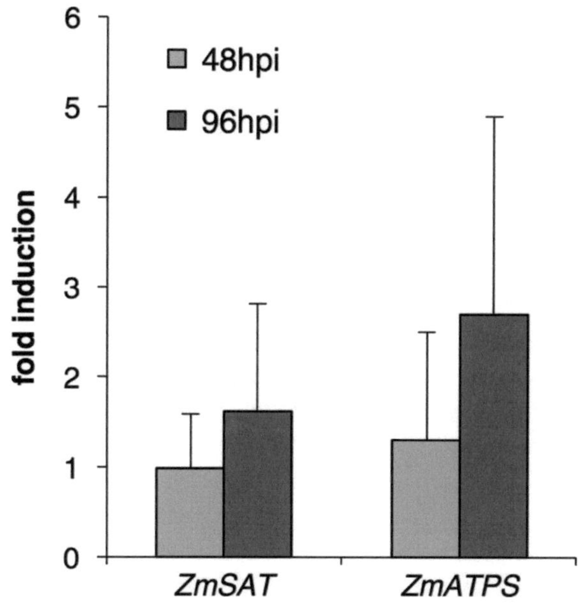

Figure 7. Expression profile of *ZmSAT* and *ZmATPS* genes in *C. graminicola*-infected roots.

4. Discussion

It has been documented that plant sRNAs can act as regulators of gene expression during plant-defence responses as reviewed in Ref. [36]. However, the mechanisms of sRNA-mediated immunity remain largely elusive, especially for host-fungi interactions. In rice cultivars that are susceptible to *Magnaporthe oryzae*, enhanced resistance could be achieved by overexpressing miR160 and miR398 [37]. MicroRNA160 targets auxin-responsive factor 16 (*ARF16*), and miR398 regulates superoxide dismutase 2 (*SOD2*), both known defence-related genes. In wheat, *B. graminis* infection was demonstrated to lead to massive adaptations of the miRNA expression profile, where miRNAs only induced in either resistant or susceptible cultivars where identified [19]. Various sRNA expression studies upon fungal infections point towards a role as fine-tuners in the concert of setting up efficient and targeted antifungal defences, rather than having direct defensive impacts [38]. In this regard, it is not surprising that sRNAs were identified as regulators of basal immunity and *R*-gene-mediated resistance. In cotton, bioinformatic approaches revealed over 300 NBS-LRR genes potentially controlled by the miR482 family [39], which cleave NBS-LRR transcripts, resulting in the generation of secondary siRNAs that even enhance the silencing of multiple NBS-LRR genes. *V. dahliae* infection leads to a down-regulation of miR482, hence to a de-repression of *R*-genes.

Notably, plant RNAi pathway components were shown in specific cases to be important for mounting proper antifungal-defence responses. RDR6-deficient plants were found to be more susceptible to *Verticillium* spp. but not to *Botrytis cinerea, Alternaria brassicicola* or *Plectosphaerella cucumerina* [40]. Similarly, *ago4* mutants were discovered to be more susceptible to *B. cinerea* and *P. cucumerina* [41, 42], possibly due to the over-induction of the SA-defence pathway which leads to diminished JA-defence responses that are important in controlling necrotrophic pathogens.

The present study widens the understanding of the putative role of sRNAs in fine-tuning plant-hormonal pathways during fungal infection. First of all, Arabidopsis sRNA pathway components were demonstrated to be required for antifungal responses against *C. higginsianum*. HYL1- and HEN1-deficient plants were more susceptible than the wild type. The higher susceptibility was accompanied by a de-regulated hormonal response. The sRNA mutants *hen1-1* and *hyl1-2* exhibited higher SA-, JA- and ABA-induction levels. The hormonal imbalance might explain the altered susceptibility to *C. higginsianum*, as enhanced SA is known to be important during biotrophic infections, whereas high JA levels are typical for defence against necrotrophs [43]. On the other hand, mutation of *RDR6* did not affect the susceptibility against *C. higginsianum* suggesting that the tasiRNA (trans-acting siRNA) pathway is not involved in antifungal responses. Recent studies demonstrated that RDR6-deficient plants were more resistant to the hemibiotrophic pathogen *Pseudomonas syringae DC3000*, presumably by a constitutive activation of the SA-dependent-defence pathway. Hence, it was speculated that RDR6 acts as a negative regulator of PTI and basal defence in Arabidopsis [44]. Notably, the hormonal imbalance discovered in the sRNA mutants could only partially explain the altered susceptibility, as higher JA levels were found in *hyl1-2* mutants, which would lead to an enhanced resistance during the necrotrophic stage of *C. higginsianum*. This suggests that sRNAs act as putative-defence coordinators beyond hormonal pathways. Consequently, the metabolomic analysis uncovered additional layers of sRNA-regulated antifungal responses, namely the proper induction of defence-related secondary metabolites. Especially *hen1-1* and *hyl1-2* mutants were found to exert a massively altered defence metabolome, and to a lesser extent also the analysed *rdr6-15* mutants. sRNAs are known to be directly involved in the regulation of secondary metabolites; overexpression of miR393 for instance was shown to increase levels of glucosinolates and decreases camalexin [45], which indicates that miR393 is involved in the re-direction of the metabolic flows. Similarly, a possible link between miR163 and the biosynthesis of secondary metabolites was described [46]. Loss or overexpression of miR163 alters the transcription of target genes and the profiles of secondary metabolites after induction by the fungal elicitor alamethicin. On the other hand, *rdr6-15* mutants showed a wild-type-like metabolomic profile in both infected and control-treated conditions, despite the levels of some compounds being slightly different in the mutant after infection. The minor differences in the metabolomic profile between *rdr6-15* and wild-type plants might be explained by the complex redundancies between the members of these protein families [3, 47]. Notably, some sRNA mutants such as *hen1-1* exhibit developmental defects, thus the genetic and metabolomic phenotype observed in response to fungi might be significantly affected by developmental pathways. However, this issue underlies all genetic studies using knock-out mutants with severe phenotypes. Using rigid statistical criteria for compound clustering, it is possible to partially differentiate developmental from antifungal responses, as shown in the PCA analysis of infected and control *hen1-1* mutants.

To extend the view on antifungal responses possibly linked to sRNA pathways, the miRNA transcriptome of the agricultural important model crop *Z. mays* infected with *C. graminicola* was analysed. During this interaction, maize was found to set-up an organ-specific miRNA profile. In the locally and systemically induced fungal-specific miRNAs, only a few were found to target defence genes. In particular, zma-miR1432, which targets a para-hydroxy-benzoate-polyprenyl transferase, was found to be up-regulated during both biotrophic and necrotrophic infection stages in maize leaves and roots. The miRNA target is essential in terpenoid-quinone synthesis. Hence, it could be speculated that its down-regulation could divert the flow of secondary metabolites from terpenoids towards flavonoid biosynthesis. This would be coherent with the fact that terpenoids play only minor roles during *C. graminicola* infection in maize [48]. The second identified miRNA linked to defence pathways was zma-miR169. This miRNA is down-regulated in response to fungal root infections, and it putatively targets a gene encoding an allene oxide synthase (*AOS*). AOS is a key enzyme in JA synthesis, thus it can be speculated that zma-miR169 acts as a suppressor of JA signalling under non-stressed conditions, whereas the down-regulation of zma-miR169 during fungal infection could promote JA synthesis. The enhanced JA levels in some Arabidopsis miRNA mutants support this hypothesis. Another yet intriguing altered miRNA was zma-miR395, which was down-regulated in *C. graminicola*-infected maize roots and systemic leaves upon root infection. The down-regulation of zma-miR395 was accompanied by the up-regulation of two of its putative targets in roots, one of them encoding an ATP sulphurylase. APS plays an important role in sulphate assimilation and glutathione synthesis; inhibiting glutathione synthesis in Arabidopsis was shown to trigger the suppression of miR395 [49], thus mimicking fungal infection. It can be speculated that the down-regulation of zma-miR395 positively regulates sulphate-mediated defence and/or the glutathione pathway. Intriguingly, miR168 that targets AGO1 was induced upon leaf infection, consistent with recent work demonstrating a similar fold induction of miR168 in Arabidopsis treated with elicitors of *F. oxysporum* [50]. Thereby, a majority of elicitor-responsive miRNAs were shown to be associated with development and miRNA homeostasis [50], corroborating the observation that sRNA pathways likely do not regulate direct-defence pathways.

Although some sRNA pathways components were shown here to be required for battling *C. higginsianum*, they seem to act as fine-tuner of defence schemes rather than to directly regulate defence genes and defensive compounds. A common picture found for sRNA mutants exhibiting higher susceptibility to *C. higginsianum* was rather a hormonal and metabolomic imbalance. Moreover, the absence of altered miRNAs targeting direct-defence genes in maize suggested an indirect defensive role of sRNAs against *Colletotrichum* spp.

5. RNA silencing and plant defence: an outlook

Altogether, the strategic outline of Arabidopsis and maize antifungal defence against *Colletotrichum* spp. points towards the concept that sRNAs are acting as fine-tuners mediating the balance of multiple genetic and metabolomic-defence layers. The sRNA-orchestrated fine-tuning of defensive layouts may provide a genetic flexibility allowing rapid and efficient adaptation of immune pathways. The question whether sRNA pathways are indispensable and pivotal antifungal-defence regulators remains debatable. Despite various studies showing the altered susceptibility of sRNA mutants, the general trend is that the outcome of sRNA-mediated defence strongly depends on a specific pathosystem. This chapter adds further highlights to this picture by showing that no miRNAs targeting classical defence pathways are de-regulated upon *C. graminicola* infection in maize, and Arabidopsis sRNA mutants under fungal attack appear to have altered metabolomic profile compared to the wild-type situation.

Nevertheless, considering the fact that sRNA pathways are also involved in setting up proper abiotic stress responses, it might represent a multi-valuable biotechnological approach to generate crops that are more efficient and variant in expressing their sRNA repertoire. Over the past years, a transgene-based approach where pathogen-targeting sRNAs are expressed in host species was repeatedly confirmed to efficiently control fungal diseases. This host-induced gene-silencing (HIGS) approach was successfully applied to a broader range of host-pathogen systems, thus bearing a valuable industrial potential. Significant drawbacks with this technology are the restrictive acceptance of genetically modified crops, and the yet elusive question of how fast pathogens evolve tolerance or resistance. For instance, *F. graminearum* sRNA mutants are showing normal virulence in wheat infection assays [51]. A yet elusive question is the role of plant endogenous sRNAs in targeting the genes of their fungal parasites. *B. cinerea* has been demonstrated to hijack plant genes using sRNA effectors [52]; thus it could be possible that plant sRNA effectors are able to infiltrate fungal cells to act as antimicrobial molecules.

Recent studies from two different research groups demonstrate fungal control by exogenous application of sRNAs to *F. graminearum* [53] and *B. cinerea* [54]. The so-called spray-induced gene silencing (SIGS) might provide novel biotechnological opportunities to control fungal diseases. Although the data from both studies are promising, it remains elusive how efficient sRNAs are compared to classical biologicals, and how broadly this technology can be applied. For instance, Botrytis is one of the few fungal species known to require a functional sRNA machinery for proper infection, hence possibly representing a special situation in sRNA-mediated plant-pathogen interaction. Moreover, exogenous control by sRNAs was efficient in controlling fungal growth on vegetables and fruits, and not demonstrated on leaves [54], suggesting efficacy only in a very specific infection condition. Altogether, it remains to be elucidated how efficient SIGS could work in field conditions, and in particular also the application spectrum of this technology. So far, this new technology lacks confirmation by additional independent studies to allow fully evaluating its industrial potential.

Prospective investigations will help in further elucidating of the full potential of sRNA-mediated antifungal defence. While the data presented here and in recent studies suggest that sRNAs are subtle players in the concert of mounted antifungal defence, and new approaches using exogenously applied sRNAs are promising, there remains challenging basic research to be completed first in order to truly understand sRNA trafficking and signalling in plant-pathogen interactions.

Acknowledgements

We are grateful to Gaétan Glauser and Armelle Vallat-Michel (Chemical Analytical Service of the University of Neuchâtel) for their technical assistance in metabolomics and hormone analysis. Financial support from the National Centre of Competence in Research 'Plant Survival' and grant number 31003A-120197, both research programmes of the Swiss National Science Foundation, is gratefully acknowledged.

Author details

Andrea Balmer[1], Emanuele De Paoli[2], Azeddine Si-Ammour[3], Brigitte Mauch-Mani[1] and Dirk Balmer[1,4]*

*Address all correspondence to: dirk.balmer@syngenta.com

1 Institute of Biology, Laboratory of Molecular and Cell Biology, University of Neuchâtel, Neuchâtel, Switzerland

2 Dipartimento di Scienze Agrarie e Ambientali, Università degli Studi di Udine, Udine, Italy

3 Research and Innovation Centre—Fondazione Edmund Mach, Michele all'Adige, Trentino, Italy

4 Syngenta Crop Protection AG, Stein, Switzerland

References

[1] Ding SW, Voinnet O. Antiviral immunity directed by small RNAs. Cell. 2007;**130**:413-426

[2] Pumplin N, Voinnet O. RNA silencing suppression by plant pathogens: defence, counter-defence and counter-counter-defence. Nature Review Microbiology. 2013;**11**:745-760

[3] Borges F, Martienssen RA. The expanding world of small RNAs in plants. Nature Reviews Molecular Cell Biology. 2015;**16**:727-741

[4] Voinnet O. Origin, biogenesis, and activity of plant microRNAs. Cell. 2009;**136**:669-687

[5] Mallory AC, Vaucheret H. Functions of microRNAs and related small RNAs in plants. Nature Genetics. 2006;**38**(Suppl):S31-S36

[6] Kruszka K, Pieczynski M, Windels D, Bielewicz D, Jarmolowski A, Szweykowska-Kulinska Z, et al. Role of microRNAs and other sRNAs of plants in their changing environments. Journal of Plant Physiology. 2012;**169**:1664-1672

[7] Baulcombe D. RNA silencing in plants. Nature. 2004;**431**:356-363

[8] Ding SW. RNA-based antiviral immunity. Nature Reviews Immunology. 2010;**10**:632-644

[9] Li Y, Lu J, Han Y, Fan X, Ding SW. RNA interference functions as an antiviral immunity mechanism in mammals. Science. 2013;**342**:231-234

[10] Chisholm ST, Coaker G, Day B, Staskawicz BJ. Host-microbe interactions: Shaping the evolution of the plant immune response. Cell. 2006;**124**:803-814

[11] Jones JD, Dangl JL. The plant immune system. Nature. 2006;**444**:323-329

[12] Pastor V, Balmer A, Gamir J, Flors V, Mauch-Mani B. Preparing to fight back: generation and storage of priming compounds. Frontiers in Plant Science. 2014;**5**:295

[13] Jin H. Endogenous small RNAs and antibacterial immunity in plants. FEBS Letters. 2008;**582**:2679-2684

[14] Padmanabhan C, Zhang X, Jin H. Host small RNAs are big contributors to plant innate immunity. Current Opinion in Plant Biology. 2009;**12**:465-472

[15] Zhang W, Gao S, Zhou X, Chellappan P, Chen Z, Zhang X, et al. Bacteria-responsive microRNAs regulate plant innate immunity by modulating plant hormone networks. Plant Molecular Biology. 2011;**75**:93-105

[16] Boccara M, Sarazin A, Thiebeauld O, Jay F, Voinnet O, Navarro L, et al. The Arabidopsis miR472-RDR6 silencing pathway modulates PAMP- and effector-triggered immunity through the post-transcriptional control of disease resistance genes. PLoS Pathogens. 2014;**10**:e1003883

[17] Navarro L, Dunoyer P, Jay F, Arnold B, Dharmasiri N, Estelle M, et al. A plant miRNA contributes to antibacterial resistance by repressing auxin signaling. Science. 2006;**312**:436-439

[18] Radwan O, Liu Y, Clough SJ. Transcriptional analysis of soybean root response to Fusarium virguliforme, the causal agent of sudden death syndrome. Molecular Plant-Microbe Interactions. 2011;**24**:958-72

[19] Xin M, Wang Y, Yao Y, Xie C, Peng H, Ni Z, et al. Diverse set of microRNAs are responsive to powdery mildew infection and heat stress in wheat (Triticum aestivum L.). BMC Plant Biology. 2010;**10**:123

[20] Yin Z, Li Y, Han X, Shen F. Genome-wide profiling of miRNAs and other small non-coding RNAs in the Verticillium dahliae-inoculated cotton roots. PLoS One. 2012;**7**:e35765

[21] Lu S, Sun YH, Amerson H, Chiang VL. MicroRNAs in loblolly pine (Pinus taeda L.) and their association with fusiform rust gall development. Plant Journal. 2007;**51**:1077-1098

[22] Guo N, Ye WW, Wu XL, Shen DY, Wang YC, Xing H, et al. Microarray profiling reveals microRNAs involving soybean resistance to Phytophthora sojae. Genome. 2011;**54**:954-958

[23] Nowara D, Gay A, Lacomme C, Shaw J, Ridout C, Douchkov D, et al. HIGS: host-induced gene silencing in the obligate biotrophic fungal pathogen Blumeria graminis. Plant Cell. 2010;**22**:3130-3141

[24] Koch A, Kumar N, Weber L, Keller H, Imani J, Kogel KH. Host-induced gene silencing of cytochrome P450 lanosterol C14alpha-demethylase-encoding genes confers strong resistance to Fusarium species. Proceedings of the National Academy of Sciences United States of America. 2013;**110**:19324-19329

[25] O'Connell RJ, Thon MR, Hacquard S, Amyotte SG, Kleemann J, Torres MF, et al. Lifestyle transitions in plant pathogenic Colletotrichum fungi deciphered by genome and transcriptome analyses. Nature Genetics. 2012;**44**:1060-1065

[26] O'Connell R, Herbert C, Sreenivasaprasad S, Khatib M, Esquerre-Tugaye MT, Dumas B. A novel Arabidopsis-Colletotrichum pathosystem for the molecular dissection of plant-fungal interactions. Molecular Plant-Microbe Interactions. 2004;**17**:272-282

[27] Huser A, Takahara H, Schmalenbach W, O'Connell R. Discovery of pathogenicity genes in the crucifer anthracnose fungus Colletotrichum higginsianum, using random insertional mutagenesis. Molecular Plant-Microbe Interactions. 2009;**22**:143-156

[28] Martins J, Branco F, Gonzalez Castro M, Cota M, Barroso J. A new concept of 3D DCS interface application for industrial production console operators. Universal Access in the Information Society. 2015;**14**:399-413

[29] Vargas WA, Martin JM, Rech GE, Rivera LP, Benito EP, Diaz-Minguez JM, et al. Plant defense mechanisms are activated during biotrophic and necrotrophic development of Colletotricum graminicola in maize. Plant Physiology. 2012;**158**:1342-1358

[30] Thomma BP, Penninckx IA, Broekaert WF, Cammue BP. The complexity of disease signaling in Arabidopsis. Current Opinion in Immunology. 2001;**13**:63-68

[31] Narusaka Y, Narusaka M, Park P, Kubo Y, Hirayama T, Seki M, et al. RCH1, a locus in Arabidopsis that confers resistance to the hemibiotrophic fungal pathogen Colletotrichum higginsianum. Molecular Plant-Microbe Interactions. 2004;**17**:749-62

[32] Balmer D, de Papajewski DV, Planchamp C, Glauser G, Mauch-Mani B. Induced resistance in maize is based on organ-specific defence responses. Plant Journal. 2013;**74**:213-225

[33] Si-Ammour A, Windels D, Arn-Bouldoires E, Kutter C, Ailhas J, Meins F, Jr., et al. miR393 and secondary siRNAs regulate expression of the TIR1/AFB2 auxin receptor clade and auxin-related development of Arabidopsis leaves. Plant Physiology. 2011;**157**:683-691

[34] Jiang H, Wong WH. SeqMap: mapping massive amount of oligonucleotides to the genome. Bioinformatics. 2008;**24**:2395-2396

[35] Balmer D, Flors V, Glauser G, Mauch-Mani B. Metabolomics of cereals under biotic stress: Current knowledge and techniques. Frontiers in Plant Science. 2013;**4**:82

[36] Weiberg A, Wang M, Bellinger M, Jin H. Small RNAs: A new paradigm in plant-microbe interactions. Annual Reviews in Phytopathology. 2014;**52**:495-516

[37] Li Y, Lu YG, Shi Y, Wu L, Xu YJ, Huang F, et al. Multiple rice microRNAs are involved in immunity against the blast fungus Magnaporthe oryzae. Plant Physiology. 2014; **164**:1077-1092

[38] Balmer D, Mauch-Mani B. Small yet mighty—microRNAs in plant-microbe interactions. Microrna. 2013;**2**:72-79

[39] Zhu QH, Fan L, Liu Y, Xu H, Llewellyn D, Wilson I. miR482 regulation of NBS-LRR defense genes during fungal pathogen infection in cotton. PLoS One. 2013;**8**:e84390

[40] Laender A, Goncalves M, Cota R, Barbosa F. Transitioning from the ecological fieldwork to an online repository: A digital library solution and evaluation. International Journal on Digital Libraries. 2007;**7**:109-112

[41] Lopez A, Ramirez V, Garcia-Andrade J, Flors V, Vera P. The RNA silencing enzyme RNA polymerase v is required for plant immunity. PLoS Genetics. 2011;**7**:e1002434

[42] Nonfiction-grades K-3. Library Talk. [Book Review]. **11**:29

[43] Pieterse CM, Leon-Reyes A, Van der Ent S, Van Wees SC. Networking by small-molecule hormones in plant immunity. Nature Chemical Biology. 2009;**5**:308-316

[44] Boccara M, Sarazin A, Thiebeauld O, Jay F, Voinnet O, Navarro L, et al. Correction: The Arabidopsis miR472-RDR6 silencing pathway modulates PAMP- and effector-triggered immunity through the post-transcriptional control of disease resistance genes. PLoS Pathogens. 2015;**11**:e1004814

[45] Robert-Seilaniantz A, MacLean D, Jikumaru Y, Hill L, Yamaguchi S, Kamiya Y, et al. The microRNA miR393 re-directs secondary metabolite biosynthesis away from camalexin and towards glucosinolates. Plant Journal. 2011;**67**:218-231

[46] Ng DW, Zhang C, Miller M, Palmer G, Whiteley M, Tholl D, et al. cis- and trans-Regulation of miR163 and target genes confers natural variation of secondary metabolites in two Arabidopsis species and their allopolyploids. Plant Cell. 2011;**23**:1729-1740

[47] Mallory A, Vaucheret H. Form, function, and regulation of ARGONAUTE proteins. Plant Cell. 2010;**22**:3879-3889

[48] Balmer D, Planchamp C, Mauch-Mani B. On the move: Induced resistance in monocots. Journal of Experimental Botany. 2013;**64**:1249-1261

[49] Nonfiction--Grades 1-3. Library Talk. [Book Review]. **11**:30

[50] Baldrich P, Kakar K, Sire C, Moreno AB, Berger A, Garcia-Chapa M, et al. Small RNA profiling reveals regulation of Arabidopsis miR168 and heterochromatic siRNA415 in response to fungal elicitors. BMC Genomics. 2014;**15**:1083

[51] Chen Y, Gao Q, Huang M, Liu Y, Liu Z, Liu X, et al. Characterization of RNA silencing components in the plant pathogenic fungus Fusarium graminearum. Scientific Reports. 2015;**5**:12500

[52] Weiberg A, Wang M, Lin FM, Zhao H, Zhang Z, Kaloshian I, et al. Fungal small RNAs suppress plant immunity by hijacking host RNA interference pathways. Science. 2013;**342**:118-123

[53] Koch A, Biedenkopf D, Furch A, Weber L, Rossbach O, Abdellatef E, et al. An RNAi-based control of fusarium graminearum infections through spraying of long dsRNAs involves a plant passage and is controlled by the fungal silencing machinery. PLoS Pathogens. 2016;**12**:e1005901

[54] Wang M, Weiberg A, Lin FM, Thomma BP, Huang HD, Jin H. Bidirectional cross-kingdom RNAi and fungal uptake of external RNAs confer plant protection. Nature Plants. 2016;**2**:16151

Chapter 3

The Use of Gamma Irradiation in Plant Mutation Breeding

Ramazan Beyaz and Mustafa Yildiz

Additional information is available at the end of the chapter

http://dx.doi.org/10.5772/intechopen.69974

Abstract

In plant breeding programs, one of the oldest methods is mutation breeding. Currently, mutation breeding has became popular among the breeders and scientists again with its use in plant biotechnology and due to some restrictions on the other techniques such as hybridization, cross breeding, and transgenic plants. Physical mutagens (X-rays, UV light, neutrons-alpha-beta particles, fast and thermal neutrons, especially gamma rays) are used more widely than chemical (ethyl methanesulfonate [EMS]) ones to artificially induce mutations (mutagenesis). However, among the physical mutagens, gamma-rays are widely used. During the irradiation of the seeds with ionizing radiation to generate mutants with desirable traits, reactive oxygen species (ROS) or free radicals can generate in cells. Although, these radicals/species generally can be very dangerous for the cell compartments, they can take an important role as a signal molecule activation of genes of antioxidant enzymes and proline, which are defense systems against these radicals in plant cells. In this chapter, usability of gamma-irradiation to provide the permanent gene expression of antioxidant enzymes and proline through the production of reactive oxygen species (ROS) is discussed.

Keywords: gamma rays, plant mutation breeding, reactive oxygen species, gene expression, antioxidant enzymes

1. Introduction

As one of the earliest agricultural activities, plant breeding has begun in very ancient times (early on Neolithic revolution, 10,000 BC) parallel to human culture [1, 2]. For the plant breeding, genetic variations are the pre-request. Mutations (natural process that creates new variants [alleles] of genes) are the main source of all genetic variation in plants as well as in any other organisms [3, 4]. According to the historical records (an ancient book "Lulan"), the first spontaneous mutant plants (cereal crops) are found in China 2317 years ago [5, 6]. After that, a few researchers reported spontaneous variation in plants between 1590 and 1968 [6]. However, the first publications of induced mutations (through X-rays) for the plant breeding was published 89 years ago by Muller and Stadler [7, 8]. As a result of these studies, mutation breeding as a new approach was added to other plant breeding methods. Thereafter, a large amount of genetic variability has been induced by various mutagens (physical and chemical) and contributed to modern plant breeding. *Nicotiana tabacum* was the first commercial mutant cultivar (called: chlorine type), which produced by inducing mutations [6]. Induced mutations were used to improve tolerant plant varieties over the past 50 years in all over the world [8]. Currently, 3246 registered mutagenic plant varieties are there in FAO/IAEA mutant data base [9].

Increasing crop yields is a major demand for assuring food security. Mutagenesis is an important tool to improve crops and has not got any regulatory restrictions as genetically modified organisms (GMOs) [10]. Plant breeding is based on the genes. Initially, breeders select new phenotypes with valuable characters without knowing the genetic constitution. The emergence of molecular genetics is parallel to understand the details of inheritance of desirable/undesirable traits and genetically controlled, modern biotechnological breeding has paved a wide road. By using DNA recombinant technologies, the gene encoding a trait precisely manipulated to create novel phenotypes. The cloned gene in respect of the source or recipient of the genes can be transferred by breeding technology known as transgenic technology. In transgenic technology, the key step is the integration of desired foreign genes into the host plant genome. For plant transformation, there are primary tree methods such as the Agrobacterium-mediated, particle bombardment, and protoplast transformation. The Agrobacterium-mediated gene transfer method is one of the most practical and suitable method [2]. The first transgenic plant (tobacco [*N. tabacum*], which contain antibiotic resistance gene) was obtained in 1980 by Marc De Block through Agrobacterium-mediated method [11]. Breeders can transfer encoding genes of new characters into plants genome through transgenic approaches. Its precision and the betterment of a trait without changing the genetic makeup of genome in elite genotypes are the main advantages [2]. Although transgenic technology has significant achievement in improving crops and has substantial commercial value, this technology has some technical obstacles. For example, in terms of highly recalcitrant to genetic transformation and regeneration, there are many economically important plant species or elite varieties of species. In addition to the technical obstacles, there are some debates about unpredictable risks of transgenic technology on environment and food safety, even though many of these debates are baseless. However, more advanced technologies have been developed to solve these ideas [2]. On the other hand, plant breeders use mutagenesis in plant breeding programs without restrictions such as the legislative constraints, licensing costs, and societal opposition of transgenic technology [10]. Although still limited to the content of the endogenous genome, mutagenesis and high-resolution screening will supply a very good complement to recombinant DNA technologies and genetically modified organisms (GMOs) in further improved new plant forms that are better adapted to change conditions of environment and the increasing global population [1].

2. Application of gamma rays to induce mutation and *in vitro* selection

As mutations may occur spontaneously, it can induce artificially. Artificially induced mutations can be created by physical mutagens, such as X-rays, gamma rays, and neutrons, and chemical mutagens, such as ethyl methanesulfonate (EMS), in plant mutation breeding [12]. Although, various mutagens (physical and chemical) are used for the induction of mutation, physical mutagens (radiation: gamma rays and X-rays) was used more frequently as compared to chemical mutagens. However, among the physical mutagens, gamma rays (1604 improved mutants) are used more widely than X-rays (561 improved mutants). Gamma rays are ionizing radiation and used in inducing mutations in seeds and other planting materials such as cuttings, pollen, or tissue-cultured calli [4, 13]. In the early twentieth century, plant biologists discovered method that the frequency of genetic variations could increase in treated seeds with chemical compounds or radioactive rays. Mutation induction has become a powerful tool for developing new and novel plant germplasm [14]. Radiation-induced mutation is one the most widely used method to improve direct mutant varieties compared to acclimatization, selection, hybridization, which are laborious, time consuming, and also with limited genetic variation [15]. This discovery later referred as plant mutation breeding. As an *in vitro* culture techniques, *in vitro* selection can be use to obtain plant genotype tolerance to adverse environment conditions such as drought, high salinity [16]. Numerous mutant or genetic material is possible due to *in vitro* selection [17]. *In vitro* selection is the greatest method to obtain the desired traits of plant, because it has the capacity to manipulate the variation to the expected result. Both abiotic and biotic tolerant plants could be obtained from the media which contain the selection agent [16]. *In vitro* selection has been practiced for desirable traits and success has been achieved in several crop plants [18–20]. Easy application of *in vitro* selection method is one of the most important requirements for the achievement of *in vitro* selection technique and to obtain the tolerant plants. Exposed biological materials by mutagens can be reliable and easy screening in a comparatively small space, which can save time, money, and space under *in vitro* conditions when comparing the greenhouse and field [21].

Beyaz [22] reported easy and reliable *in vitro* selection protocol and optimization for the creation of irradiation-based mutagenesis in sainfoin (*Onobrychis viciifolia* Scop.). Sainfoin (*O. viciifolia*) is a significant forage legume species, which grown widespread than the other *Onobrychis* species, agriculturally [23]. The seeds of sainfoin (*O. viciifolia* Scop.) were irradiated to 0, 400, 500, and 600 Gy using an experimental ^{60}Co source (dose rate of 0.483 kGy h^{-1}) at the Sarayköy Nuclear Research and Training Center (SANAEM) of the Turkish Atomic Energy Authority (Ankara, Turkey). Irradiated and unirradiated (control) seeds of sainfoin were sowned into Murashige and Skoog (MS) medium supplemented with 150 mM NaCl (**Figure 1A**). Seeds were allowed to germinate and develop at 20 ± 1°C under cool white fluorescent light (27 μmol m^{-2} s^{-1}) with a (16 h light/8 h dark) illumination period for 10 days. After that, selected plantlets were transferred to MS medium without NaCl for continuous development for 2 weeks (**Figure 1B**). For acclimatization, following steps were applied. Advanced plantlets (**Figure 1C**) were transferred to plastic glasses (**Figure 1D**). After that, plastic glasses placed in freezer bags (**Figure 1E** and **F**) to provide suitable moisture and adaptation to external condition for 1 week grown in the plant growth chamber. Plantlets in freezer bags were transferred to plant growth chamber (**Figure 2A**) and the mouths of the freezer bags slowly open during 1 week period, and at the end of 1 week, all plantlets were removed from freezer bags (**Figure 2B**). Every 2 days for 2 weeks, acclimatized plantlets watered with 30 ml tap water containing 150 mM NaCl for *in vivo* selection. Most of the plantlets died (**Figure 2C**) during treatments of NaCl, survived plantlets were selected (**Figure 2D**). Survived plantlets (**Figure 2E**) were transferred to plastic glasses which contain soil without NaCl for more

Figure 1. Irradiated and unirradiated seeds of sainfoin in MS-basal selection media supplemented with 150 mM NaCl (A). After 10 days of seed germination, selected plantlets and advanced plantlets in MS-basal medium without NaCl (B, C). An acclimatization process of plantlets to external conditions (D–F).

Figure 2. General view of plantlets in freezer bags (A). Plantlets which removed freezer bags in plant growth chamber (B). Dead plants as a result of salt application (C). Survived plantlets (D). Survived plantlets after treatments of NaCl (E). Washing root of survived plants with tap water to get rid of NaCl and transferred to new soil without NaCl (F and G).

development (**Figure 2E** and **G**). Superior plants (putative mutants) in terms of salt tolerance were moved to a growth chamber (**Figure 3A** and **B**) for a while, and after that, they were transplanted in field (**Figure 3C–F**).

Figure 3. General view of survived plants of sainfoin in growth chamber (A and B). Transferred putative mutants against to salt stress in the field (C and D). Putative mutant sainfoin plants in the field after 1.5 months (E) and 2 months (F) (Location: The University of Ankara, Faculty of Agriculture, Department of Field Crops, Turkey, photos were taken on 21 March 2013).

3. Ionize radiation, reactive oxygen species (ROS) and defense systems of ROS

Ionizing radiation causes biological injury in exposed biological materials. The first target of ionizing radiation is water molecule, which is ubiquitous in any organisms. The cell is composed of ~80% water [24]. As a result of excitation and ionization reactions, water molecule (H_2O) and $H^•$ and OH radicals are generated [25]. Gamma rays cause to produce free radicals (free radicals like $O_2^{•-}$ and $OH^•$ and nonradicals like H_2O_2 and $^1O_2$) as known reactive oxygen species (ROS) through direct interactions of radiation with target macromolecules or via products of water radiolysis [13, 24, 26]. The formation of reactive oxygen species (ROS) occurs in the general metabolism of the plant cell.

However, such as other environmental stress, radiation lead to increase the formation of ROS in plant cell due to damage of cellular homeostasis and cause progressive oxidative damage and finally cell death [27].

Reactive oxygen species (ROS) control many different processes in plants [28]. Plants has two antioxidant machinery, one of them is antioxidative enzymes, including ascorbate peroxidase (APX), glutathione reductase (GR), superoxide dismutase (SOD), catalase (CAT), guaiacol peroxidase (GPX), monodehydroascorbate reductase (MDHAR), and dehydroascorbate reductase (DHAR). The other one is nonenzymatic antioxidants like ascorbic acid (AA), reduced glutathione (GSH), α-tocopherol, carotenoids, flavonoids, and the osmolyte proline [26]. ROS depends on ionizing radiation level that causes damage or modification of components in plants, ultimately affecting morphology, physiology, anatomy, and biochemistry of plants [13, 29]. Currently, scientific evidence shows that ROS play an important signaling role in plants and regulate biological activities such as growth, development, and especially response to biotic and abiotic stress factors [26, 27]. ROS can induce injury of cell compartment, but on the other hand, they induce new gene expression in cells [30]. However, Esnault et al. [31] hypothesized that ROS (mainly H_2O_2) can play a secondary role are in signaling process of cell. And after a first stress, plants can be more tolerant to a new stress synthesis due to secondary metabolites. Moreover, using gamma rays can create a permanent gene expression of antioxidative enzymes for scaving "oxidative stress" start from the first generation of plants. And this provides to improve superior plants varieties against biotic and abiotic stress factors.

Here in the main question: Can we use ionize radiations (especially, gamma rays) to generate mutants with desirable characteristics via supply permanent gene expression of antioxidant enzymes such as catalase (CAT), superoxide dismutase (SOD), glutathione reductase (GR), and ascorbate peroxidase (APX) and also osmoprotectants such as proline and transmissible to the progeny in plants?

Although there are lots of works related to changing transcriptional regulation of various types of genes (especially genes of antioxidative enzymes) due to gamma irradiation, there are limited reports on permanent and transmissible increased transcript levels of genes, which induced by gamma rays, of antioxidative enzymes in plants.

Beyaz et al. [32] reported that permanent production of antioxidant enzymes and proline in M1 plants of sainfoin (*Onobrychis viciifolia* Scop.) under *in vitro* conditions. In the study conducted by Beyaz et al. [32], the aim was to investigate effects of gamma radiation on physiological responses of the M1 sainfoin plants. Seeds of sainfoin ecotype 'Koçaş' were exposed to 0, 400, 500, and 600 Gy from a ^{60}Co source at a dose rate of 0.483 kGy h^{-1}. Irradiated and unirradiated seeds were sown into culture vessels containing MS-basal medium to be cultured for 30 days under *in vitro* conditions. At the end of this period, seedlings, which germinated from the radiated and unirradiated seeds, were transferred into pots in a growth chamber for 30 days more. Chlorophyll contents, activities of superoxide dismutase (SOD), catalase (CAT), ascorbate peroxidase (APX), and glutathione reductase (GR), as well as contents of malondialdehyde (MDA) and proline were examined in unirradiated and irradiated 60-day-old seedlings. Overall, the activities of the antioxidant enzymes (SOD, CAT, and GR) and contents of chlorophyll and proline in the leaves tended to increase after irradiation in a dose-dependent manner. By contrast, the activity of APX decreased. The lipid peroxidation characterized by the MDA content remained unchanged, except after irradiation to 500 Gy. The highest CAT activity and the highest proline content were observed after irradiation to the highest dose of 600 Gy. The highest SOD and GR activities were observed after irradiation to the lowest tested dose of 400 Gy. This is the first study that provided basic information on the impact of gamma radiation on physiological responses of sainfoin and its radiosensitivity. These findings will be useful in development of a mutation breeding program of sainfoin.

Also Zaka et al. [33] query the same questions in their investigation and reported that the low chronic ionizing radiation provide genetically transmissible gene expression of antioxidant enzymes such as (SOD, GR, CAT, POD, and GPDH) to the progeny of Spila capillata (Poaceae), which grown two contaminated areas (5.4 and 25 μSv h^{-1}) on the Semipalatinsk nuclear test site in Kazakhstan. They considered evolutionary point of view and answered their observation with the natural populations that can change their genetic structure under environmental constraints and facilitates adaptation. The selective pressure of low radioactive contamination levels (leading to gamma-irradiation dose rates as low as 4.5 and 25 μSv h^{-1}) may have played an important role during 50 years. However, Çelik et al. [25] reported that a high level of ascorbate peroxidase activity (APX) in the leaves generated from irradiated soybean seeds, compared to the unirradiated group. The APX activity increased in the unirradiated group over the experimental period and was increased by 3.6-fold at 14 days and 3.8-fold at 21 days after irradiation ($P \leq 0.05$). Kim et al. [34] found that gene expression of antioxidant enzymes such as superoxide dismutase (SOD), ascorbate peroxidase (APX), catalase (CAT), and peroxidase (POD) can be induced depending on increasing gamma-irradiation dosage level in Triticeae species. Plant cells change their protein metabolism under main stress conditions. Resistant or stress proteins are induced in response to gamma-irradiation stress, and this defense mechanism change patterns of gene expression, especially stress-inducible gene expression, which produces qualitative and quantitative changes in total soluble protein [34, 35]. The different gene expression pattern of antioxidant enzymes can be observed in the irradiated plant species [36].

According to Mohammed et al. [37], a substantial variation of the protein patterns of cowpea seeds occur by gamma rays induced, and this variation has occurred because of the new expression of some polypeptide, the silence of others, and overexpression of a third class polypeptides. Aly and El-Beltagi [38] showed that increase in GST, CAT, SOD, and POD activities in *Vicia faba* L. seeds could be attributed to ionizing irradiation stress. Cho et al. [39] investigated the expression patterns of diverse genes at various time points after gamma irradiation of young tobacco plants and found three different gene expression pattern (increased, decreased, and no response) of antioxidative enzymes (CAT, SOD, and GST). However, Al-Rumaih and Al-Rumaih [40] reported that gamma irradiation affected antioxidant enzyme activities in the three investigated species of wheat. Moreover, the increased activity of antioxidant enzymes (SOD, CAT, POX, APOX) in response to gamma-irradiation treatment in many plant species (Nicotiana, *Triticum aestivum*, sugar cane, *Phaseolus vulgaris*, radish, groundnut, and pepper) were reported [41–46].

As a result of metabolic pathways of aerobic cells, free radicals and ROS are occurred and influenced most of the biological activity. On the other hand, products of oxidation reactions play an important role some biological process such as aging, some degenerative diseases, differentiation, and development, including serving as subcellular messengers in gene regulatory and signal transduction pathways. Several studies reported the hypothesis. The hypothesis is that shifts oxidant/antioxidant equilibrium in cell may also affect developmental pathways in a different of tissues from phylogenetically diverse organisms [47].

Over expression of antioxidant enzyme genes likely arise due to an efficient regulatory mechanism to provide cells with resistance [33, 40]. Sen et al. [48] reported that the activities of several antioxidant enzymes were evaluated in both vegetative and flowering stages, and mutant lines (wheat irradiated by 200 Gy) showing the highest biochemical performance. These studies clearly indicated that gene expression of antioxidant enzymes can be made permanently in genome of progeny of plants by gamma-rays irradiation.

4. Conclusion

It is estimated that the population of the world will be 9.1 billion, which is compared to 34% today, in 2050. Populations of developing countries will be the biggest in this increase. Urbanization will continue at an increasing rate and the majority of the world's population (~70%) will live in urban (compared to 49% today) [49]. However, food production is not increasing parallel to the human population and a large part of the population of today's world is not already well fed. Therefore, humanity has two critical problems such as controlling population growth and increasing food production. It is important for food safety that people are economically and easily accessible to food. For the food security, physical availability and affordability of food are the most important criteria. Induced mutations have a vital role in increasing world food security, because of induced mutations provide new food crop varieties, contributed to the significant increase in crop production and supplied directly accessible of food for the locations people [8]. Mutagenesis has become widespread again in plant breeding during the last decades. Plant mutagenesis, which creates new variation in crop plants, coupled with *in vitro* selection and plant biotechnology methods allows breeders to select for characters that were tough obtain in breeding for only a few decades ago [1]. This is a viable option that increases productivities via "smart" plant varieties that can produce more yield. However, the genetic similarities among crop varieties or unvarying parental materials, which are weaker than the same stress, are limited to uncover new alleles of genes in cropping systems. Therefore, creating new combination of genes causes to break yield plateaux and enhance tolerance. Induced mutation uncovers novel alleles that are used to breed superior crop varieties [50]. The uncovering of new genetic variation in inbred elite cultivars supplies a unique possibility to characterize novel traits, while the lines protect their excellence in the agriculture. Depending on the increasing technology such as whole-genome sequencing and high-resolution analytical techniques, we accumulate more genetic information from a wide range of crop plants and also gain both money and time compared to traditional breeding techniques. However, generated markers in this process offer to the stack of the useful characters, paving the way for developing complex multigenic characters such as abiotic stress resistance [1].

As a results of biotic and abiotic stress factors, the production of reactive oxygen species (ROS) lead to increase in plant cells and cause oxidative stress. Scavenging mechanisms such as antioxidant enzymes keep plants from adverse effect of oxidative stress [30]. If we succeed supplying permanent gene expression of antioxidant enzymes and osmoprotectants such as proline and glycine-betaine by gamma radiation in the plant cell, we can provide tolerances of plants to almost adverse environmental stress factors. For the future, it seems that mutagenic crops continue their important role in plant breeding.

Author details

Ramazan Beyaz[1*] and Mustafa Yildiz[2]

*Address all correspondence to: ramazanbeyaz@gmail.com

1 Department of Soil Science and Plant Nutrition, Faculty of Agriculture, Ahi Evran University, Bagbasi, Kırsehir, Turkey

2 Department of Field Crops, Faculty of Agriculture, University of Ankara, Dıskapı, Ankara, Turkey

References

[1] Sikora P, Chawade A, Larsson M, Olsson J, Olsson O. Mutagenesis as a tool in plant genetics, functional genomics, and breeding. International Journal of Plant Genomics. 2011;**2011**:1-13. DOI: 10.1155/2011/314829

[2] Xiong J, Ding J, Li Y. Genome-editing technologies and their potential application in horticultural crop breeding. Horticulture Research. 2015;**2**:1-10. DOI: 10.1038/hortres.2015.19

[3] Kharkwal MC. A brief history of plant mutagenesis. In: Shu QY, Forster BP, Nakagawa H, editors. Plant Mutation Breeding and Biotechnology. Wallingford: CABI; 2012. pp. 21-30

[4] Oladosu Y, Rafii MY, Abdullah N, Ghazali H, Asfaliza R, Rahim HA, Miah G, Usman M. Principle and application of plant mutagenesis in crop improvement: A review. Biotechnology & Biotechnological Equipment. 2016;**30**(1):1-16. DOI: 10.1080/13102818.2015.1087333

[5] van Harten AM. Mutation Breeding: Theory and Practical Applications. 1998. Cambridge: Cambridge University Press

[6] Solanki RK, Gill RK, Verma P, Singh S. Mutation breeding in pulses: An overview. In: Khan S, Kozgar MI. Breeding of Pulse Crops, 1st ed. Ludhiana: Kalyani Publishers; 2015. pp. 85-103

[7] Pathirana R. Plant mutation breeding in agriculture. CAB Reviews Perspectives in Agriculture Veterinary Science Nutrition and Natural Resources. 2011; **6**(032):1-20. DOI: 10.1079/PAVSNNR20116032

[8] Kharkwal MC, Shu Y. The role of induced mutations in world food security. In: Shu QY, editor. Induced Plant Mutations in the Genomics Era. Rome: Food and Agriculture Organization of the United Nations; 2009. pp. 33-38

[9] Mutant Data Base. [Internet]. 2017. Available from: https://mvd.iaea.org/#!Search [Accessed: 23-05-2017]

[10] Parry MAJ, Madgwick PJ, Bayon C, Tearall K, Hernandez-Lopez A, Baudo M, Rakszegi M, Hamada W, Al-Yassin A, Ouabbou H, Labhilili M, Phillips AL. Mutation discovery for crop improvement. Journal of Experimental Botany. 2009;**60**(10):2817-2825

[11] Lijsebettens MV, Angenon G, De Block M. Transgenic plants: From first successes to future applications. International Journal of Developmental Biology. 2013;**57**:461-465. DOI: 10.1387/ijdb.130063mv

[12] Ahloowalia BS, Maluszynski M. Induced mutations—A new paradigm in plant breeding. Euphytica. 2001;**118**(2):167-173

[13] Ali H, Ghori Z, Sheikh S, Gul A. Effects of gamma radiation on crop production. In: Hakeem KR, editor. Crop Production and Global Environmental Issues. Springer International Publishing, Switzerland 2016. pp. 27-78. DOI: 10.1007/978-3-319-23162-4_2

[14] Penna S, Vitthal SB, Yadav PV. *In vitro* mutagenesis and selection in plant tissue cultures and their prospects for crop improvement. Global Science Book. 2012;**6**(1):6-14

[15] Hanafiah DS, Trikoesoemaningtyas, Yahya S, Wirnas D. Induced mutations by gamma ray irradiation to Argomulyo soybean (*Glycine max*) variety. Nusantara Bioscience. 2010;**2**(3):121-125

[16] Lestari EG. *In vitro* selection and somaclonal variation for biotic and abiotic stress. Biodiversitas. 2006;**7**(3):297-301

[17] Maluszynski M, Ahloowalia BS, Sigurbjornsson B. Application of *in vivo* and *in vitro* mutation techniques for crop improvement. Euphytica. 1995;**85**:303-315

[18] Nikam AA, Devarumath RM, Shitole MG, Ghole VS, Tawar PN, Suprasanna P. Gamma radiation, *in vitro* selection for salt (NaCl) tolerance, and characterization of mutants in sugarcane (*Saccharum officinarum* L.). In Vitro Cellular & Developmental Biology-Plant. 2014;**50**:766-776. DOI: 10.1007/s11627-014-9630-4

[19] Suhesti S, Khumaida N, Wattimena GA, Syukur M, Husni A, Hadipoentyanti E. Gamma irradiation and *in vitro* selection could increase drought tolerance in sugarcane. International Journal of Sciences: Basic and Applied Research (IJSBAR). 2015; **23**(2):370-380

[20] El-Sabagh AS, Barakat MN, Genaidy EAE. Towards *in vitro* selection studies for salinity tolerance in *Canino apricot* cultivar. Effect of gamma irradiation on *in vitro* mutation and selection for salt-tolerance. Advances in Horticultural Science. 2011;**25**(4):260-263

[21] Çiftçi CY, Türkan AD, Khawar KM, Atak M, Özcan S. Use of gamma rays to induce mutations in four pea (*Pisum sativum* L.) cultivars. Turkish Journal of Biology. 2006;**30**:29-37

[22] Beyaz R. Determination of Salt Tolerance of Different Sainfoin (*Onobrychis viciifolia* Scop.) Ecotypes and Improvement of New Sainfoin Lines via In vitro Mutagenesis Technique (PhD thesis). Ankara University; Ankara 2014

[23] Carbonero CH, Mueller-Harvey I, Brown TA, Smith L. Sainfoin (*Onobrychis viciifolia*): A beneficial forage legume. Plant Genetic Resources: Characterization and Utilization. 2011;**9**(1):70-85. DOI: 10.1017/S1479262110000328

[24] Azzam EI, Jay-Gerin JP, Pain D. Ionizing radiation-induced metabolic oxidative stress and prolonged cell injury. Cancer Letters. 2012 Feb 31:**327**(0):48-60. DOI: 10.1016/j.canlet.2011.12.012

[25] Marcu D, Damian G, Cosma C, Cristea V. Gamma radiation effects on seed germination, growth and pigment content, and ESR study of induced free radicals in maize (*Zea mays*). Journal of Biological Physics. 2013;**39**:625-634. DOI: 10.1007/s10867-013-9322-z

[26] Das K, Roychoudhury A. Reactive oxygen species (ROS) and response of antioxidants as ROS-scavengers during environmental stress in plants. Frontier in Environmental Science. 2014;**2**(53):1-13. DOI: 10.3389/fenvs.2014.00053

[27] Sharma P, Jha AB, Dubey RS, Pessarakli M. Reactive oxygen species, oxidative damage, and antioxidative defense mechanism in plants under stressful conditions. Journal of Botany. 2012; **2012**:1-26. DOI: 10.1155/2012/217037

[28] Mittler R, Vanderauwera S, Gollery M, Breusegem FV. Reactive oxygen gene network of plants. Trends in Plant Science. 2004;**9**(10):490-498. DOI: 10.1016/j.tplants.2004.08.009

[29] Çelik Ö, Atak Ç, Suludere Z. Response of soybean plants to gamma radiation: Biochemical analyses and expression patterns of trichome development. Plant Omics Journal. 2014;**7**(5): 382-391

[30] Gill SS, Tuteja N. Reactive oxygen species and antioxidant machinery in abiotic stress tolerance in crop plants. Plant Physiology and Biochemistry. 2010;**48**:909-930. DOI: 10.1016/j.plaphy.2010.08.016

[31] Esnault MA, Legue F, Chenal C. Ionizing radiation: Advances in plant response. Environmental and Experimental Botany. 2010;**68**:231-237

[32] Beyaz R, Sancak C, Yildiz Ç, Kuşvuran Ş, Yildiz M. Physiological responses of the M1 sainfoin (*Onobrychis viciifolia* Scop.) plants to gamma radiation. Applied Radiation and Isotopes. 2016;**118**:73-79. DOI: 10.1016/j.apradiso.2016.09.005

[33] Zaka R, Vandecasteele CM, Misset MT. Effect of low chronic doses on antioxidant enzymes and G PDH activities in Stipa capillate (Poaceae). Journal of Experimental Botany. 2002;**53**(376):1979-1987

[34] Kim DY, Hong MJ, Park CS, Seo YW. The effects of chronic radiation of gamma ray on protein expression and oxidative stress in *Brachypodium distachyon*. International Journal of Radiation Biology. 2015;**91**(5):407-419

[35] Corthals G, Gygi S, Aebersold R, Patterson S. Identification of proteins by mass spectrometry. In: Rabilloud T, editor. Proteome Research: Two-Dimensional Gel Electrophoresis and Identification Methods. New York: Springer; 2000. pp. 197-231

[36] Ling APK, Chia JY, Hussein S, Harun AR. Physiological responses of citrus sinensis to gamma irradiation. World Applied Sciences Journal. 2008;**5**(1):12-19

[37] Mohammed AHMA, Mohammed HI, Zaki LM, Mogazy AM. Pre-exposure to gamma rays alleviated the harmful effect of salinity on cowpea plants. Journal of Stress Phsiology and Biochemistry. 2012;**8**(4):199-217

[38] Aly AA, El-Beltagi HES. Influence of ionizing irradiation on the antioxidant enzymes of *Vicia faba* L. Grasas Y Aceites, 2010;**61**(3):288-294

[39] Cho HS, Lee HS, Pai H. Expression pattern of diverse genes in response to gamma irradiation in *Nicotiana tabacum*. Journal of Plant Physiology. 2000;**43**(2):82-87

[40] Al-Rumaih MM, Al-Rumaih MM: Influence of ionizing radiation on antioxidant enzymes in three species of Trigonella. American Journal of Environmental Sciences. 2008;**4**(2):151-156

[41] Sreedhar M, Chaturvedi A, Aparna M, Pavan Kumar D, Singhal RK, Venu-Babu P. Influence of γ-radiation stress on scavenging enzyme activity and cell ultra structure in groundnut (*Arachis hypogaea* L.). Pelagia Research Library. 2013;**4**(2):35-44

[42] Singh RK, Chandra P, Singh J, Singh DN. Effect of gamma-ray on physio-biochemical parameters of sugar cane. Journal of Nuclear Agriculture & Biology. 1993;**22**:65-69

[43] Wada H, Koshiba T, Matsui T, Sato M. Involvement of peroxidase in differential sensitivity to irradiation in seedlings of two Nicotiana species. Plant Science. 1998;**132**:109-119

[44] Lee HY, Kim JS, Baek MH, Yoo JC. Effects of low dose irradiation on physiological activities of radish (*Raphanus sativus*) during early growth and reduction of UVB stress. Journal of Korean Society for Horticultural Science. 2003;**44**:314-320

[45] Stoeva N. Physiological effects of the synthetic growth regulator Thidiazurol (drop) on gamma-irradiated stress in peas plants (*Pisum sativum*). Journal of Central European Agriculture. 2002;**6**(2):349-358

[46] Kim JH, Baek MH, Chung BY, Wi SG, Kim JS. Alterations in the photosynthic pigments and antioxidant machineries of red pepper (*Capsicum annuum* L.) seedlings from gamma-irradiated seeds. Journal of Plant Biotechnology. 2004;**47**(2):314-321

[47] Allen RG. Oxidative stress and superoxide dismutase in development, aging gene and regulation. Age. 1998;**21**:47-76

[48] Sen A, Ozturk I, Yaycili O, Alikamanoglu S. Drought tolerance in irradiated wheat mutants studied by genetic and biochemical markers. Journal of Plant Growth Regulation. 2016. **2017**:1-11 DOI: 10.1007/s00344-017-9668-8

[49] How to Feed the World in 2050. [Internet]. 1999. Available from: http://www.fao.org/fileadmin/templates/wsfs/docs/expert_paper/How_to_Feed_the_World_in_2050.pdf. [Accessed: 22-05-2017]

[50] Chikelu Mba. Induced mutations unleash the potentials of plant genetic resources for food and agriculture. Agronomy. 2013;**3**:200-231. DOI: 10.3390/agronomy3010200

Section 3

Ways to Accelerate the Plant Generation Time

Chapter 4

Hd3a Florigen Recruits Different Proteins to Reveal Its Function in Plant Growth and Development

Yekti Asih Purwestri, Febri Adi Susanto and Hiroyuki Tsuji

Additional information is available at the end of the chapter

http://dx.doi.org/10.5772/intechopen.70263

Abstract

The nature of Hd3a protein in rice and its ortholog FT in *Arabidopsis* as a florigen has been proposed. However, molecular mechanism of its function still remains to be investigated. Therefore, it is important to search their interaction partners to better understand their signaling in flowering. As a long-distance signal that moves along leaf cells and the vascular system of leaves and stem and exerts its action in apical buds, it is important to determine the possible mediators of such common responses activated by Hd3a. To search Hd3a interactor, yeast two-hybrid screening have performed by using a cDNA library. A wide range of Hd3a interacting proteins involved in signaling were identified, including GF14c, OsKANADI and the BRI1 kinase domain interacting protein 116b (BIP116b). To reveal its function, Hd3a recruits different protein in plant developmental stage. It is possible that Hd3a and its partner(s) may form a platform for cross-talk between signal transduction pathways. Another homolog of Hd3a in many plants was identified and sugessted that Hd3a/FT has versatile role in plant development. This role depend on its partner and interaction to achieve its function. Our understanding in floral transition in rice would make for better crop management in future.

Keywords: Hd3a, florigen, interacting proteins

1. Introduction

Transition from vegetative phase to flowering involves many genetic pathways that interact with the external signals, such as day length and temperature, and internal signals such as hormones and developmental controls. One of the most important factors controlling flowering plants is response to daylight or photoperiod [1]. Based on photoperiodism, two model plants, *Arabidopsis thaliana* and *Oryza sativa* are used to study on regulation of flowering time of long-day plant (LDP) and short-day plant (SDP), respectively. Three genes, which constitute a major genetic pathway in the photoperiodic regulation of flowering in rice, have recently been isolated. *O. sativa GIGANTEA* (*OsGI*), an ortholog of *Arabidopsis GI*, *Heading date 1* (*Hd1*), an ortholog of *Arabidopsis CO* (*CONSTANS*), and *Heading date 3a* (*Hd3a*), an ortholog of *FLOWERING LOCUS T* (*Arabidopsis FT*) are shown to form the main pathway for photoperiodic regulation of flowering in rice. These three genes were conserved between rice and *Arabidopsis*; however, the differences in their regulation results in either SDP or LDP (**Figure 1**). The major difference between rice, a SDP, and *Arabidopsis*, a LDP, was shown to be the regulation of *Hd3a/FT* by *Hd1/CO*. Under LD conditions, this regulation is positive in *Arabidopsis* while negative in rice [2].

In *Arabidopsis*, *GI* encodes a nuclear protein [3]. Expression of *GI* mRNA exhibits a circadian rhythm and *gi* mutants defects in clock function, indicating that *GI* is closely associated with the clock itself. *CO* encodes a transcription factor with B-box type zinc fingers thought to mediate protein-protein interaction. *FT* encodes a protein with homology to Raf kinase inhibitor protein (RKIP). It is a powerful promoter of flowering, activating the floral meristem identity gene, *APETALA1* (*AP1*) and is the target of several pathways controlling flowering time [4].

Hd3a was initially identified by quantitative trait locus (QTL) promoting flowering under short day-length condition. Using map-based cloning, *Hd3a* was determined as an ortholog of *FT* in *Arabidopsis* [5]. *Hd3a* overexpression under 35S promoter, native promoter, and vascular-specific promoters showed early flowering phenotype [6]. On the other hand, the suppression of *Hd3a* by RNA interference exhibited delayed flowering in rice [7]. These results strongly suggest that Hd3a protein function for flowering promotion.

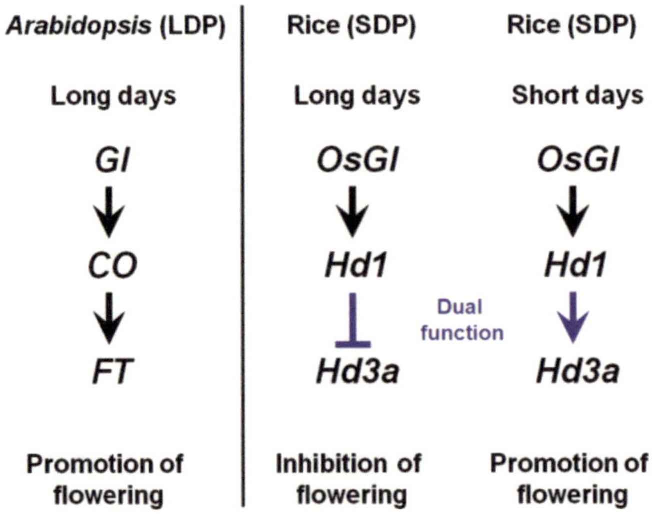

Figure 1. Flowering pathway regulation in *Arabidopsis* and rice.

http://dx.doi.org/10.5772/intechopen.70263

Hd3a shares sequence similarity with the mammalian phosphatidylethanolamine-binding protein (PEBP or RAF1 kinase inhibitor protein—RKIP) (**Figure 2**). The PEBP family regulates signaling pathways to control growth and differentiation. The PEBPs seem to act biochemically as inhibitors, binding signaling components to modulate the flux through their pathways.

The crystal structure of PEBP from human and bovine sources, CEN protein from *Antirrhinum* [8] and that of the Terminal Flowering Locus (TFL) and FT proteins from *Arabidopsis* [9] have been determined. In **Figure 3**, the molecular model of Hd3a protein was built using the Swiss-Prot automated comparative protein modeling server, based on its sequence homology to two members of the RKIP protein family whose structures have been determined by X-ray crystallographic methods (*Arabidopsis* FT and TFL1, protein databank accession numbers 1WKP and 1WKO, respectively). Structural analysis of these proteins indicated the accessibility of the ligand-binding pocket to interact with the proteins partner.

PEBPs might also act as either scaffolds for or regulators of signaling complexes, as showed by the finding that Self-Pruning (SP) and Single-Flower Truss (SFT), as a tomato homolog of Terminal Flowering Locus 1 (TFL1) and Flowering Locus T (FT), respectively, can interact with a range of diverse proteins [10, 11]. Several studies of protein interactions involving FT/Hd3a orthologs have been published. In *Arabidopsis*, FT interacts with the basic/leucine zipper (bZIP) transcriptional factor FD [12, 13]. Interestingly, the bZIP transcription factor Self-pruning G-box protein (SPGB) in tomato, a homolog of FD, interact with SP and SFT as well. Moreover, FT interacts with FD and 14-3-3 proteins (**Table 1**) [11–13]. However, no Hd3a interacting proteins have yet been identified in rice.

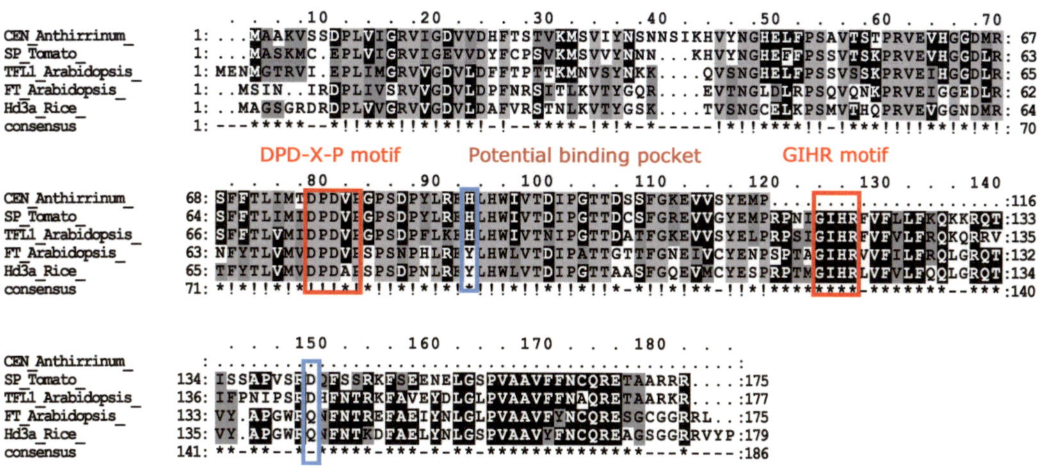

Figure 2. Hd3a shares high homology with other phosphatidylethanolamine binding protein or Raf kinase inhibitor protein (PEBP/RKIP) in various plant species. Hd3a and FT has 73% homology. The blue box indicates the amino acid difference which is responsible for its function as flower inducer (tyrosin in FT and Hd3a) or repressor (histidine in CEN, SP, and TFL1).

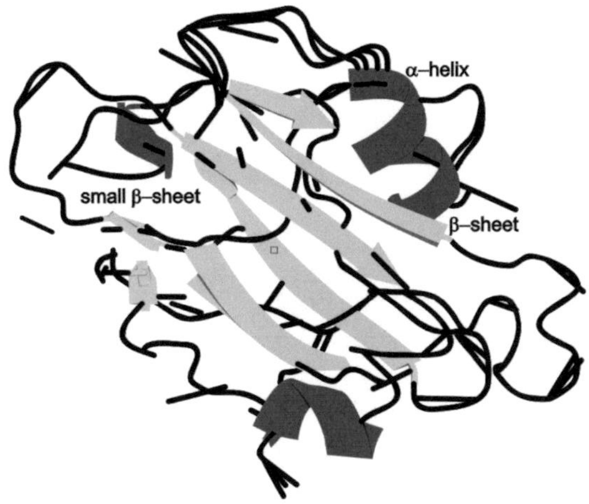

Figure 3. Molecular model of Hd3a. Hd3a protein contains a large central β-sheet (yellow ribbons), which is flanked on one side by a smaller β-sheet and on the other by an α-helix (red ribbons).

In *Arabidopsis*, it is believed that the combination of interacting proteins, resolved crystal structures, and mutant phenotype analysis will lead to a comprehensive understanding of the mechanisms that facilitate the switch from vegetative phase to reproductive phase. It seems likely that Hd3a/FT is involved not only in flowering, but also in other aspects of growth and development in plant architectures [11, 15–18]. This function will be achieved by interacting with its partners. Hd3a might recruit different proteins to perform its roles in plant growth and development, particularly during floral transition.

In this chapter, we will discuss the regulation of Hd3a florigen in rice and the identification of novel interaction partners for rice Hd3a protein using yeast two-hybrid screening. The interaction between Hd3a and its partners was further confirmed by several methods, such as yeast two-hybrid assay using full-length cDNA, *in vitro* pull-down assay, co-immunoprecipitation, and bimolecular fluorescence complementation (BiFC). The expression pattern and subcellular localization of each Hd3a interacting partner provided important insights into its function. To further characterize the function of Hd3a interacting proteins in plant growth

Protein	**Interacting proteins**	**References**
Mammalian PEBP	Raf1	[14]
Self-Pruning (SP) tomato	14-3-3 (adapter protein) SPGB (a putative bZIP transcription factor) SPAK (SP-associated kinase)	[10]
Flowering Locus T (FT) *Arabidopsis thaliana*	FD (bZIP transcription factor)	[12, 13]
Single-Flower Truss (SFT) tomato	SPGB 14-3-3/74	[11]

Table 1. PEBP/RKIPs interacting proteins.

and development, particularly during the floral transition, a reverse genetics approach was used to generate gain of function mutants (overexpression)/knockdown mutants using RNAi or knockout mutants.

2. Flowering regulation in rice

In rice (*O. sativa* L.), flowering is mainly induced by photoperiod. *GIGANTEA* in rice (*OsGI*) is a clock-regulated gene, which was first identified by the differential display method, and its expression is high in the middle of day [2]. Under inductive SD conditions, *OsGI* promotes the expression of *Heading date 1* (*Hd1*), and *Hd1* activates *Hd3a* expression. However, under noninductive LD conditions, *Hd1* suppresses *Hd3a* expression [2]. *OsGI* acts as the primary upstream regulator of *Hd1* expression [19]. OsGI is a large protein that is present in both the nucleus and cytoplasm of rice cells [12]. Suppression of *OsGI* by RNAi or antisense expression caused late flowering and reduced *Hd1* transcription under SD conditions [2].

Several unique genes in rice were isolated. *Early heading date1* (*Ehd1*) encodes a B-type response regulator and is a unique flowering time gene [20]. *Ehd1* promotes floral transition preferentially under SD conditions, even in the absence of functional alleles of *Hd1*. Expression analysis revealed that *Ehd1* functions upstream of *Hd3a, RFT1,* and some MADS-box genes [20]. More recently, *Ghd7* (for grain number, plant height, and heading date 7), which encodes a *CO, CO-LIKE,* and *TIMING OF CAB1* (*CCT*)-domain protein, was isolated from natural variants in rice [21]. *Ghd7* affects transcript levels of *Ehd1* and *Hd3a* under LD conditions and delayed the flowering. In the same condition, *Ghd7* does not affect *Hd1*. Therefore, rice have different flowering pathway: *Hd1* pathway and *Ehd1* pathway. These two different pathway integrated with environmental condition, in this case is photoperiod and regulate the FT-like genes to initiate or delay the flowering [22]. Another unique flowering gene in rice, *RID/OsId1/ehd2*, yields an extremely late flowering phenotype under both SD and LD conditions. This gene encodes a putative transcription factor with a zinc finger motif, and is an ortholog of *INDETERMINATE1* (*ID1*), which promotes flowering in maize (*Zea mays*). Specifically, it promotes the floral transition, mainly by upregulating *Ehd1* and genes downstream of *Ehd1*, such as *Hd3a* and *RFT1* [23–25].

Hd3a expression under SD conditions is also regulated by phytochrome. In the *se5* (the photoperiodic sensitivity 5) mutant, which lacks a functional gene encoding heme oxygenase [26], an enzyme that is required for loss-of-function alleles in one of the three rice phytochromes [27], rice exhibited early in flowering. *Hd1* expression is not affected by *se5* or *phyB* mutations; thus, phytochrome represses *Hd3a* expression downstream or independently of *Hd1* expression under SD conditions.

During the vegetative phase in rice, the shoot apical meristem (SAM) produces a series of leaves. The vegetative parts of the rice plant, consisting of root, culm, and leaves, form a tiller. A dramatic change occurs during the transition from vegetative to reproductive stages, with the tiller terminating to produce leaf and the panicle (inflorescence) being generated on the uppermost internode of the culm (**Figure 4**). The development stage of the young panicle is also related to the timing of leaf emergence. The differentiation stage of the young panicle almost

Figure 4. Rice plant during the developmental stages: vegetative phase, early reproductive phase (left panel) and during seed maturation (right panel).

directly correlates with the start of emergence of the fourth leaf (counted downward from the flag leaf). At the time of flag leaf (small last leaf) emergence, the glumes flower primordial has already differentiated and pollen mother cells are being formed [28]. The flag leaf, contributes largely to the filling of grains because it supplies photosynthetic products mainly to the panicle.

Flowering time in rice is indicated by the emergence of the flag leaf or the panicle (heading date). The panicle is initiated when the first bract primordium differentiates on the shoot apex, approximately 30 days before panicle emergence (heading). The start of differentiation of the glumes flower primordial follows after the rachis-branches differentiation has occurred (24 days before heading). The pistil and stamen differentiate 20 days before heading. Meiosis in the anther occurs 12 days before heading, and flower organ completion occurs 1–2 days before heading.

3. Yeast two-hybrid screening for searching Hd3a-interacting proteins

An important element in the characterization of Hd3a function is the identification of other proteins with which it interacts. To reveal the function of Hd3a in rice, yeast two-hybrid system was used as a tool for screening Hd3a-interacting proteins. This system consist of a fusion of protein interest to a DNA-binding domain (DBD-Hd3a) as a bait, and a fusion of cDNA library to a transcriptional activation domain (AD-cDNA library) as a prey (**Figure 5**). The DNA binding domain (DBD) recognizes a specific sequence in the DNA upstream of a promoter and the activation domain (AD) stimulates transcription by binding to RNA polymerase. If the two domains interact, they will activate transcription. In the two-hybrid system, both the DBD domain and the AD domain are fused to two other proteins. If the bait captures the prey, means that proteins are interacting, a complex will be formed and the reporter gene will be activated. The reporter gene is used to monitor for a successful interaction.

As a tool for searching Hd3a interactors, the construction of a bait and prey is important. cDNA library were constructed from leaf blades harvested 35 days after sowing at ZT 0, 2, and 4 when transition from vegetative to reproductive phase occurred and *Hd3a* was highly expressed [7]. The Hd3a full-length cDNA was constructed as a bait and expressed in yeast system (**Figure 6**).

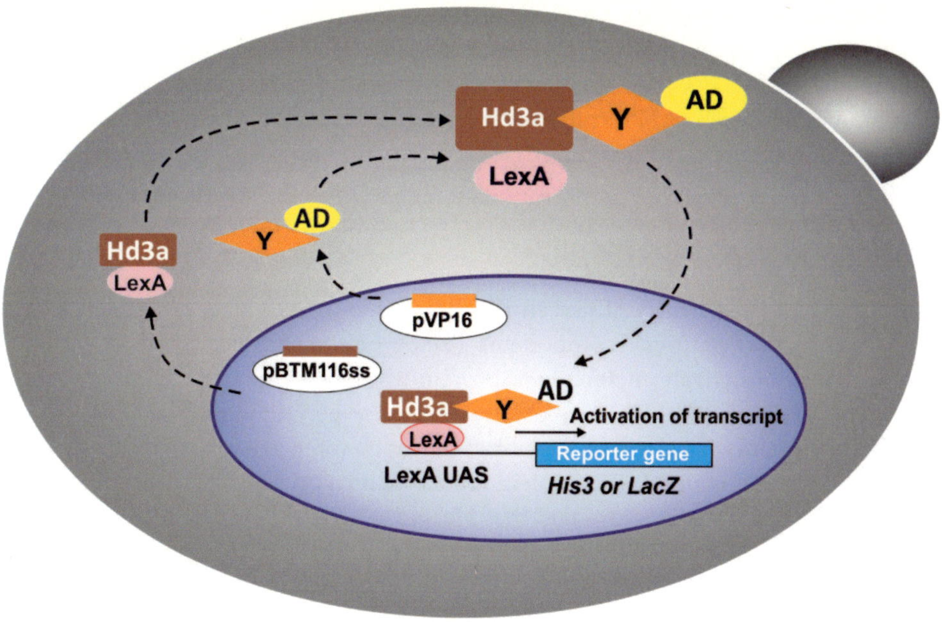

Figure 5. Yeast two-hybrid system. Positive interactions can be detected by selecting on plates lacking histidine, followed by screen for β-galactosidase expression.

For the large-scale yeast two-hybrid screen, a total of ~1.6×10^6 transformants were screened for activation of histidine synthase 3 (HIS3) onto selective medium plates without amino acids leucine (L), tryptophan (W) and histidine (SC-His). In the first screening on media SC-His, 354 colonies were obtained. The next screening on medium SC-His containing 3-AT, a total of 96 colonies' unique proteins were identified as positive clones. To eliminate false positive, we used 2.5 mM 3-AT, a competitive inhibitor of the *His3* gene product (histidine synthase),

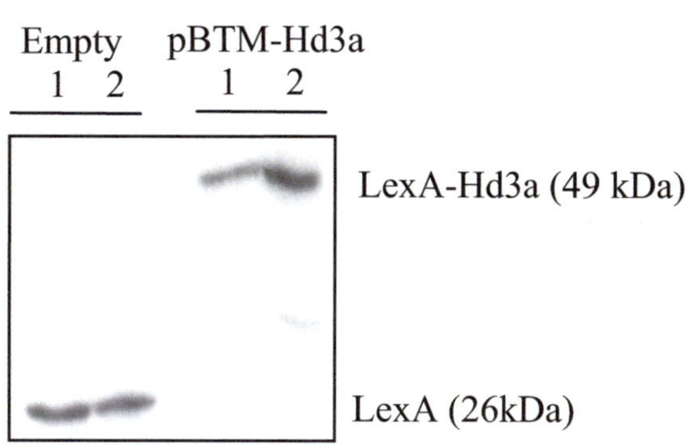

Figure 6. The Hd3a protein expression (pBTM-Hd3a) in yeast strain L40. The protein was run on 10% SDS-PAGE and electrotransferred onto an immobilon-P membrane followed by detection of Hd3a-LexA fusion using the anti-LexA antibody. Molecular weight of Hd3a-Lex A and Lex A are 49 and 26 kDa, respectively.

which is the reporter gene for the interaction in the yeast two-hybrid system. The 96 positive clones with the size ranged between 600-2000 bp then were checked to confirm the presence of "in frame" cDNA-AD fusion. Some clones were found in full-length of cDNA, but the other clones were not. They lack of N-terminal coding region. The clones were then retransformed into yeast containing Hd3a bait to confirm their interaction on selective medium SC-His plus 3-AT (**Figure 7**). The flowchart of yeast two hybrid screens is presented in (**Figure 8**). For the result, a diverse range of interactor proteins can be identified in this yeast two hybrid system (**Table 2**).

The main group of proteins identified in this yeast two-hybrid screen belongs to the class of signal transduction pathway components (57%). Others are classified as proteins that are involved in carbohydrate metabolism, protein/RNA/DNA synthesis, and proteins with unknown function. Interestingly, our yeast two-hybrid screening identified a diverse range of proteins that are mainly involved in signaling. In this chapter, we focus on three candidates: (i) GF14c (G box factor 14c), a 14-3-3 protein identified as a protein that is involved in signaling pathway. The recent results revealed that Ha3a, 14-3-3 and FD form a hexameric florigen activation complex (FAC) and in shoot apical meristem cells, 14-3-3 protein act as intracellular receptor for florigen [29]; (ii) BRI1-kinase domain (KD)-interacting protein 116b, and (iii) a novel myb transcription factor-like protein, namely OsKANADI1.

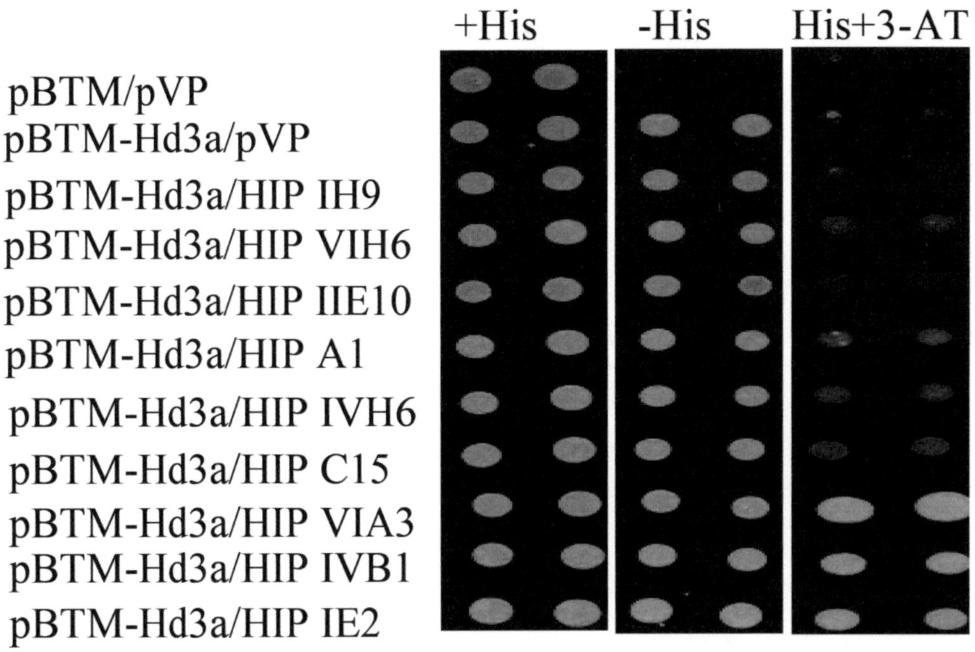

Figure 7. Interaction of Hd3a with its partners in yeast two-hybrid system. The growth of yeast colonies on the plate (LWH) lacking leucine (L), tryptophan (W), and histidine and with 2.5 mM 3-AT (3-aminotriazole) indicates a positive interaction between Hd3a and the particular Hd3a-interacting proteins. 3-AT is a competitive inhibitor of the *HIS3* gene product (histidine synthase), which is the reporter gene for the interaction in the yeast two-hybrid system. Each clone of Hd3a interacting protein (HIP) was spotted onto selective plate.

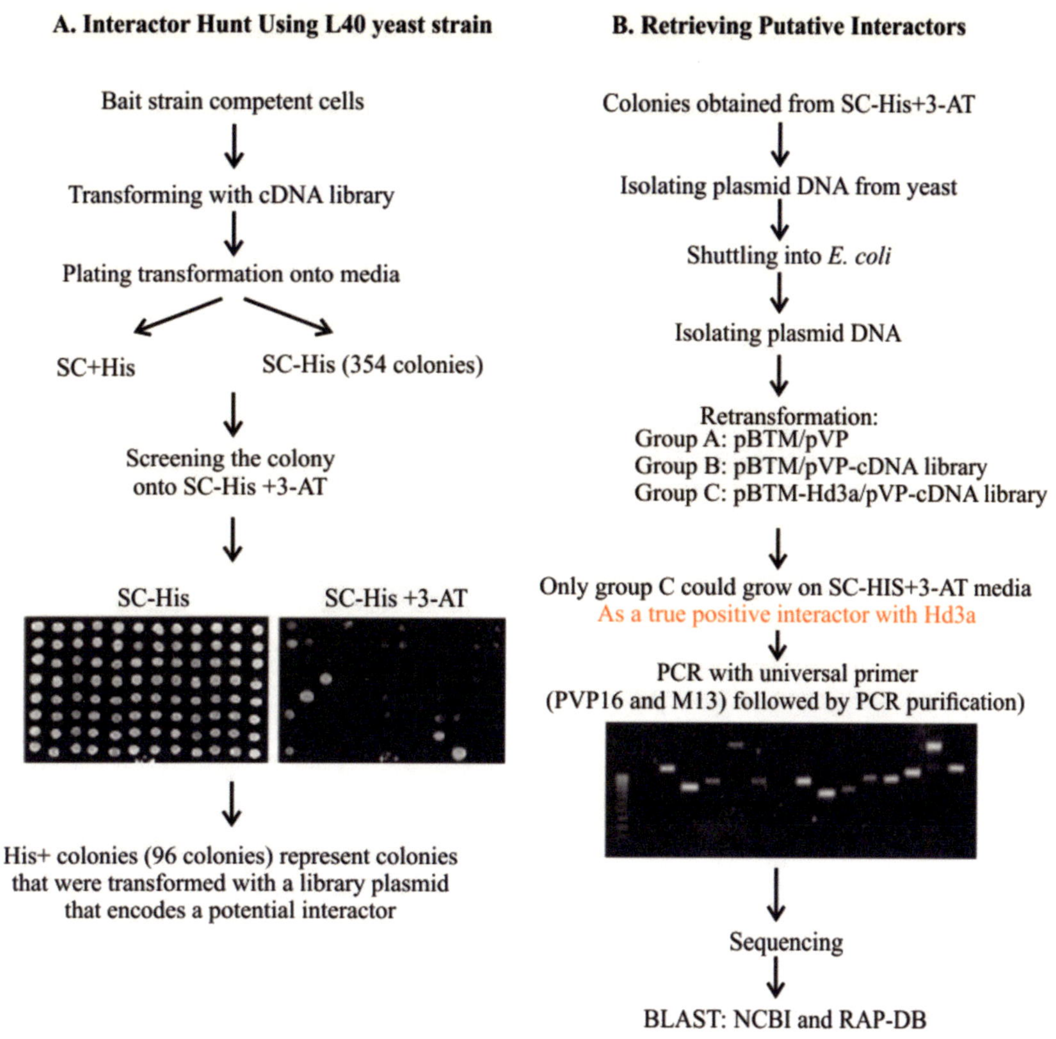

Figure 8. Flowchart of yeast two-hybrid screening. (A) Interactor hunt using L40 yeast strain; (B) Retrieving putative interactors.

Homologous proteins	Accession number	Two-hybrid*	Amino acids**	Domain/Function
Carbohydrate metabolism				
Ribulose 1,5-biphosphate carboxilase (RuBisCO) small subunit	gi|3063524	150	175	RuBisCO
RuBisCO activase large isoform precursor protein	gi|77552725	162	466	RuBisCO
Protein synthesis/DNA and RNA binding proteins				
Putative RNA helicase	gi|34897325	127	756	–
Putative 40S ribosomal protein	gi|50939279	211	305	–

Homologous proteins	Accession number	Two-hybrid*	Amino acids**	Domain/Function
Translation initiation factor 5A/eIF5A	gi|50918889	161	161	–
Linker histone	gi|50921921	185	188	–
Signaling				
GF14c (G-box factor 14-3-3c protein)	**gi|50903393 AK122149/Os8g0430500**	**115**	**256**	**14-3-3, adaptor, chaperone, scaffolding**
BRI1-KD interacting protein 116b/ BIP116b	**–**	**112**	**366**	**TPX2 domain**
OsKANADI1	**gi|41053181 Os02g0696900**	**147**	**361**	**GARP domain**
Beta cyanoalanine synthase	gi|55233175	377	377	–
Putative endonuclease/phosphatase	gi|37532222 AK100514/Os10g0203000	209	483	phosphatase
Putative anthranilate phosphoribosyltransferase	gi|50905035	121	141	–
Putative NADPH-dependent retinol dehydrogenase/reductase	gi|50726535	170	253	–
Putative SF16 protein	gi|50931997	240	500	–
Putative heat shock 70 KD protein (HSP70)	gi|27476086	157	657	chaperone
Ser/Thr kinase	gi|34894358	205	461	kinase
Putative glutathione S-transferase	gi|50900303	142	223	–
Thioredoxin H-type (TRX-H)	gi|50934769	204	350	–
Unknown				
Unknown protein	gi|50938743	68	68	–
Unknown protein	gi|54287657	91	91	–

* Length in amino acids identified from yeast two-hybrid screening.

** Length in amino acids of protein.

Table 2. Proteins identified by yeast two-hybrid screening.

4. Rice flowering regulation by Hd3a and GF14c interaction

14-3-3 proteins in general bind to phosphoserine-modified proteins, as well as to some non-phosphorylated proteins such as exoenzyme S, which has no phosphorylated residue in its binding motif. They regulate the activities of a wide array of targets via direct protein-protein interactions, and effect changes in the client proteins. These changes can vary from inactivation or activation of the enzymatic activity of a target protein, to degradation or protection from degradation of the target, to movement of the target from one cellular location to another (usually nuclear-cytoplasmic shuttling). Plant 14-3-3 proteins bind a range of TFs and other signaling proteins, and have pivotal roles in regulating developmental and stress responses.

To reveal the biological functions of 14-3-3 in plants, the common approaches of generating overexpression, RNA suppression, or knockout plant lines probably have limited potential, for two reasons. First, because they interact with so many different targets, multiple pleiotropic effects would be expected. Second, 14-3-3s are encoded by a gene family with at least 12 expressed members in *Arabidopsis* [30], 8 members in rice [31, 32], and similar numbers in other plants, such as tomato and tobacco, which adds the complicating issues of isoform specificity and redundancy. However, working with a family of proteins that play so many roles will be interesting, given the challenge of trying to identify a particular protein-protein interaction associated with a phenotype using 14-3-3 mutants (overexpression or knockout mutants). Rather than focusing on 14-3-3s themselves, more targeted and informative approaches would be to identify specific interactions using biochemical or yeast two-hybrid methods, followed by *in vivo* confirmation and directed investigation of the potential 14-3-3 binding sites in target proteins.

Several studies of protein interactions involving FT/Hd3a orthologs have been published. In *Arabidopsis*, FT interacts with FD and 14-3-3 proteins [12, 13]. There are several SP interacting proteins in tomato, including a 14-3-3 family member, protein kinase and bZIP transcription factor [10]. SFT, another tomato ortholog of FT/Hd3a, also interacts with 14-3-3 as well as bZIP [11]. In rice, GF14c (G-box factor 14-3-3c protein) has been identified as a partner of Hd3a to activate particular gene in certain pathway. Since the 14-3-3 family members known to interact with various protein, the role of GF14c in Hd3a signaling is remarkable to elucidate. Functional analysis using knockout and overexpression approach were performed and the result indicate that *GF14c* is a negative regulator by interacting with *Hd3a* [33]. No diurnal changes or developmental patterns of *GF14c* expression were observed, indicating that *GF14c* is expressed independently of the photoperiod and abundantly throughout plant development.

4.1. Subcellular localization of Hd3a and GF14c

To identify the intracellular localization of Hd3a and GF14c, a fusion construct to express mCherry fluorescent protein-linked Hd3a under the ubiquitin promoter, and GFP-linked GF14c driven by the CaMV 35S promoter were made. These constructs were introduced into rice protoplasts. In all of the rice protoplasts observed in this experiment, Hd3a-mCherry localized in both cytoplasm and nucleus; however, GF14c-GFP was predominantly visualized in the cytoplasm (75% of rice protoplasts observed) (**Figure 8**). The predominant cytoplasmic localization of 14-3-3 proteins has led to the hypothesis that they might act as cytoplasmic anchors that either block import into the nucleus or other organelles, or promote export from organelles into the cytoplasm [34].

4.2. *In vitro* and *in vivo* interaction of Hd3a and GF14c

The interaction between Hd3a and GF14c was confirmed using several methods, including a GST pull-down assay, yeast two-hybrid and a co-immunoprecipitation [33]. A GST-Hd3a fusion protein was pulled down with His-tagged GF14c, indicating that Hd3a interacted with GF14c *in vitro*. Results of this experiment were thus consistent with the results of the yeast two-hybrid experiment. An *in vivo* interaction was also demonstrated by a co-immunoprecipitation experiment using rice suspension culture cells overexpressing myc-tagged Hd3a.

4.3. Bimolecular fluorescence complementation (BiFC) of Hd3a and GF14c

To determine the distribution of Hd3a and GF14c *in vivo*, the BiFC technique was performed. The expression vector of Hd3a and GF14c fused to the N-terminal half of mVenus (Vn) and C-terminal half of mVenus (Vc), respectively. Those vectors were then transiently expressed on rice cell protoplasts. Plasmid mChery was used as a marker for transformed cells. The transformed rice protoplasts showed an interaction between Hd3a and GF14c as observed by strong green fluorescence that concentrated in the cytoplasm. The combination treatment of expression vectors using GUS to confirm the interaction cleary showed that the venus fluorescence only observed up to 10% of the transformed cells. This, indicated Hd3a only interact with GF14c mainly in cytoplasm [33].

4.4. Confirmation of Hd3a interacting protein GF14c by mutant generation

Several methods has been used to confirm the interaction between Hd3a and GF14c. All the methods used (yeast-two hybrid, in vitro pull-down, and BiFC assays) clearly showed that Hd3a and GF14c have strong interaction in rice. The subcellular distribution of Hd3a as well as GF14c demonstrated their localization in both the cytoplasm and nucleus.

GF14c acts as negative regulator of flowering and the overexpression of GF14c caused late flowering phenotype. Since the expression of Hd3a was observed both in the cytoplasm and nucleus, whereas GF14c is mainly in the cytoplasm, the possible interaction of these two protein is interesting to study. The mechanism of interaction of which GF14c would inhibit shuttling of Hd3a from the cytoplasm into the nucleus were confirmed by BiFC experiment. Based on data obtained it is clearly showed that Hd3a-GF14c is localized in cytoplasm. This indicate Hd3a-GF14c interaction lead to Hd3a cyoplasmic retention. The increasing of cytoplasmic retention of Hd3a can be explained as a model of late flowering by overexpression of GF14c. An example in *Arabidopsis*, the floral transition initiation in SAM, FT interacts with FD (bZIP transcription factor) that localized in the nucleus to induce the expression of target gene such as *AP1* [12]. In rice, this process could be attenuated by the cytoplasmic retention of Hd3a by GF14c. Several lines of evidence indicate a function for 14-3-3 proteins in nuclear-cytoplasmic shuttling in the signal transduction pathway [35, 36].

Another possible mechanism which can be explain the phenotype of GF14c overexpression and knockout mutant lines is the interaction between GF14c and Hd3a inhibit the movement of Hd3a from leaf to SAM. When GF14c expression level is low, Hd3a protein is capable to move along the floem to the SAM, since HD3a is small protein (about 20kDa) that is below of the size limit of plasmodesmata [37, 38]. The *gf14c* knockout mutant exhibited early flowering compare to WT. Remarkably, this mutant shown another phenotype such as dwarfism and increased tiller number. These pehotype suggesting that GF14c has another function independenlty with respect of flower induction at SAM.

According to Taoka [29] finding, 14-3-3 protein acts as intercellular receptor for Hd3a in the shoot apical cells. When Hd3a protein is transported to SAM, it will interact with 14-3-3 and form florigen activation complex (FAC). This FAC complex will be translocated into nucleus and binds to *OsFD1* (*FD* homolog of *A. thaliana*). The expression of *OsFD1* will lead to activation

http://dx.doi.org/10.5772/intechopen.70263

of flowering identity genes, resulting in the flowering. This finding helps us to understand the flowering mechanism in rice and also offers new approaches to manipulate the plants flowering in future.

5. Hd3a interaction with BIP116b and OsKANADI

Analysis of Hd3a interacting proteins in yeast two-hybrid screening identified a novel protein homolog of BIP116 (brassinosteroid-insensitive 1 [BRI1] kinase domain interacting protein 116) and a novel putative transcription factor belonging to the KANADI domain protein family, namely OsKANADI1 (OsKAN1).

5.1. Hd3a interact with BIP116b

BRI1 has an extracellular domain (containing an N-terminal signal peptide, and LRR-RK and island domains), a transmembrane domain, a kinase domain and a C-terminal peptide [39]. Only a partial C-terminal region fragment was identified initially from yeast two-hybrid screening. An experiment using full-length Hd3a and full-length BIP116b, either as bait or as prey, further confirmed that Hd3a and BIP116b interact in the yeast system. To further study the interaction between Hd3a and BIP116b *in vitro*, a GST pull-down assay was performed. A GST-BIP116b fusion protein was pulled down with His-tagged Hd3a. However, an interaction could not be found between Hd3a and BIP116b, even though both proteins were expressed in *Escherichia coli*. One possible explanation for this discrepancy is that posttranslational modification of BIP116b may be necessary for binding.

The transgenic plant for either overexpression or RNAi mutant were generated. The results showed that the BIP116ab RNAi plants exhibited delayed flowering compared to wild-type plant, and the overexpressed plant showed branching and increasing tiller number similar to Hd3a overexpressed plants. Therefore, further characterization in mutant plants will be interesting to get an insight into its function.

5.2. Hd3a interact with OsKANADI1

In *Arabidopsis*, *KANADI* genes function in lateral polarity in organs including roots, leaves, and flowers. *KANADI* is required for abaxial identity in both leaves and carpels. It encodes a nuclear-localized protein in the GARP family of putative transcription factors [40]. GARP homologs constitute a large family of DNA-binding proteins in plants that may be needed for a variety of key cellular functions including regulation of transcription, phosphor transfer signaling, and differentiation. A GARP motif was also found in the identified *KANADI* product [41].

KANADI acts antagonistically to the *class III HZ-Zip* genes. *KANADI* genes are expressed in a pattern complementary to that of the *class III HD-Zip* genes in the shoot; *KANADI* expression occurs in the phloem and abaxial regions of lateral organs early in development [41–43]. While *KANADI* genes do not appear to be required for proper meristem function, they are needed for pattern formation of organs produced by the shoot apical and vascular meristems.

Sequence analysis showed that OsKAN1 consists of six exons. Phylogenetic and comparative genetic analysis indicated that OsKAN1 is an ortholog of AtKANs and is most closely related to AtKAN4. The predicted OsKAN1 polypeptide contains a potential phosphorylation site in the C-terminal region which may be important for interaction with other proteins involved in signaling. The overall identity between *KANADI* family members is low, but they are nearly identical within the GARP domain. The sequence similarity found in this subset of GARP genes may indicate that they have overlapping or partially redundant functions.

Suppression of OsKANADI1 by RNAi displayed branching and increasing tiller number in several lines [44]. This phenotype resembles to the Hd3a overexpressed plants indicating that they possibly function in similar pathway. Further characterization using the OsKANADI mutants plants in the next generation remains to be investigated.

6. Hd3a homolog in plant development

The role of Hd3a as flowering mobile signal in rice and *Arabidopsis* leads to investigation of its function in another plants. The diverse role of FT was identified in many plants. FT protein involved in vernalization, stomatal opening, dormancy and tuberization [45–48]. Navarro et al. [49] provided a new insight in understanding the florigen as graft-transmissible signal and designated its role as tuberi. Hd3a homolog were overexpressed in potato under noninductive long-day condition and resulting in tuberization and flower development. A grafting experiment using tobacco florigen as scion and potato as rootstock to test the hypothesis confirmed that FT proteins were transmitted from scion to rootstock and induced tuberization. This experiment suggests that FT/Hd3a is a strong tuberization inducer.

The flowering and tuberization pathway are regulated in different pathways, indicating that FT has many homolog in potato. This hypothesis was confirmed by Navarro et al. [49]. In potato, there are three homolog of FT. StSP3A regulate the tuber formation, StSP3D regulate the flowering, and StSP5G as the repressor of tuberization. Expression of *StSP6A* gene is strong in leaves and stolon of short-day induced plants. Interestingly, 30-minute night break was reported to inhibit tuberization. The *StSP3D* expression is necessary for flowering induction. However, this gene is not required for normal tuberization. Interestingly, the StSP5G is an antagonist for StSP6A. The expression of StSP5G was high in long-day condition [47]. The role of Hd3a/FT protein in potato flowering and tuberization is well understood. This photoperiodic pathway is conserved in potato, rice, and *Arabidopsis*.

Recent study in onion tuberization reveals that its induction is controlled by Hd3a/FT homolog in the photoperiodic pathway. Onion is a biennial plant that forms tuber during the summer and flower after the winter in the next summer. There are many Hd3a/FT homologs found in onion and the pattern is the same with potato. There are Hd3a/FT homologs that regulate the tuber/bulb formation (AcFT1 and AcFT4) and flowering (AcFT2) [50]. As seen in potato tuberization, AcFT4 in onion has the same role as StSP5G in potato that inhibits onion tuber formation; while AcFT1 is a strong inducer for tuberization (tuberigen) under long-day condition. This finding suggests that Hd3a/FT has a versatile role in plant development.

Tuberigen and florigen are known to be mobile/graft-transmissible signals. Several experiments showed that this signal is interchangeable between species in Solanaceae family (potato/tobacco, and potato/tomato) and can induce tuberization of potato and onion, but cannot in other plant species. As example, the ability to form a tuber is limited in potato species among Solanaceae family. Due to those fact, there must be an unknown mechanism present in the tuberization process in potato. The underlying mechanism that involves FT roles needs further investigation.

Another Hd3a role in rice as mobile signal that induced branching was reported by Tsuji et al. [18]. Their findings suggest that Hd3a was moved along the xylem to promote branching in rice lateral meristem. The formation of florigen activation complex (FAC) for the developmental process other than flowering is not well understood. The experiment proved that Hd3a protein was transported to lateral meristem in the lateral bud and its transport is required for branching. This finding suggests that Hd3a has the function besides flowering.

Our work in local black rice namely "Cempo Ireng" which has very long harvest period also reveal that Hd3a gene is conserved. The regulation of flowering pathway of rice in short- and long-day condition is well understood. However, when rice are planted in neutral day condition such as tropical condition, the photoperiodic pathway might be affected. Our result suggests that Hd3a, which normally regulate flowering under short-day condition, is expressed in neutral day condition earlier compared to its homolog RFT1, which is responsible for flowering induction in long-day condition. In "Cempo Ireng," two flowering genes are expressed. The expression pattern of both genes are redundant to induce the flowering under neutral day condition. RFT1 and Hd3a are known to have different induction pathways. Our result suggests that RFT1 and Hd3a may have independent pathway in rice flowering regulation under neutral day condition according to their expression pattern (data not shown). The Hd3a/RFT1 interactor of black rice that involved in flowering induction is of interest for future study to understand the flowering mechanism under neutral day condition and shortening the harvest period of black rice "Cempo Ireng."

7. Conclusion and future prospects

As the floral stimulus that controls floral transition in the SAM, Hd3a has the capacity to traffic from cell to cell and move long distances via the phloem. To reveal its function, Hd3a recruits different proteins in plant developmental stage. A wide range of proteins that interact with Hd3a were identified using yeast two-hybrid screening. GF14c (a 14-3-3 homolog), OsKANADI (a novel transcription factor containing a GARP domain), and the BRI1 kinase domain-interacting protein 116b (BIP116b) are the Hd3a interactors of interest. It is possible that Hd3a and its partner(s) may form a platform for cross-talk between signal transduction pathways. Another homolog of Hd3a in many plants was identified and it was suggested that Hd3a/FT has versatile role in plant development. This role depends on its partner and interaction to gain its function. Further study using reverse genetics to obtain mutants, either gain-of-function mutants or suppression mutants by RNA interference of Hd3a partners to get insight into the function in plant growth and development, particularly during floral transition is important. Our understanding in floral transition in rice would make for better crop management in future and represent an important addition to our knowledge about FT signaling in plants.

Acknowledgements

This work was partially supported by The Ministry of Research, Technology and Higher education of Republic of Indonesia under the University Priority Research Program 2016 to YAP. We thank Badan Penerbit dan Publikasi Universitas Gadjah Mada (BPP UGM) for supporting this publication.

Author details

Yekti Asih Purwestri[1,2]*, Febri Adi Susanto[2] and Hiroyuki Tsuji[3]

*Address all correspondence to: yekti@ugm.ac.id

1 Faculty of Biology, Universitas Gadjah Mada, Yogyakarta, Indonesia

2 Research Center for Biotechnology, Universitas Gadjah Mada, Yogyakarta, Indonesia

3 Kihara Institute for Biological Research, Yokohama City University, Yokohama, Japan

References

[1] Imaizumi T, Kay S. Photoperiodic control of flowering: Not only by coincidence. Trends in Plant Science. 2006;11:550-558

[2] Hayama R, Yokoi S, Tamaki S, Yano M, Shimamoto K. Adaptation of photoperiodic control pathways produces short-day flowering in rice. Nature. 2003;422:719-722

[3] Huq E, Tepperman JM, Quail PH. GIGANTEA is a nuclear protein involved in phytochrome signaling in *Arabidopsis*. Proceedings of the National Academy of Sciences of the United States of America. 2000;97:9789-9794

[4] Simpson GG. Evolution of flowering in response to day length: Flipping the *CONSTANS* switch. BioEssays. 2003;25:1-4

[5] Kojima S, Takahashi Y, Kobayashi Y, Monna L, Sasaki T, Araki T, Yano M. *Hd3a,* a rice ortholog of the *Arabidopsis FT* gene, promotes transition to flowering downstream of *Hd1* under short-day conditions. Plant and Cell Physiology. 2002;43:1096-1105

[6] Tamaki S, Matsuo S, Wong H, Yokoi S, Shimamoto K. Hd3a protein is a mobile flowering signal in rice. Science. 2007;316:1033-1036

[7] Komiya R, Ikegami A, Tamaki S, Yokoi S, Shimamoto K. Hd3a and RFT1 are essential for flowering in rice. Development. 2008;135:767-774

[8] Banfield M, Brady R. The stucture of *Antirrhinum* centroradialis protein (CEN) suggests a role as a kinase regulator. Journal of Molecular Biology. 2000;297:1159-1170

[9] Ahn J, Miller S, Winter V, Banfield M, Lee J, Yoo S, et al. A divergent external loop confers antagonistic activity on floral regulators FT and TFL1. EMBO J. 2006;25:605-614

[10] Pnueli L, Gutfinger T, Hareven D, Ben-Naim O, Ron N, Adir N, et al. Tomato SP-interacting proteins define a conserved signaling system that regulates shoot architecture and flowering. Plant Cell. 2001;13:2687-2702

[11] Lifschitz E, Eviatar T, Rozman A, Shalit A, Goldshmidt A, Amsellem Z, et al. The tomato FT ortholog triggers systemic signals that regulate growth and flowering and substitute for diverse environmental stimuli. Proceedings of the National Academy of Sciences of the United States of America. 2006;103:6398-6403

[12] Abe M, Kobayashi Y, Yamamoto S, Daimon Y, Yamaguchi A, Ikeda Y, et al. FD, a bZIP protein mediating signals from the floral pathway integrator FT at the shoot apex. Science. 2005;309:1052-1056

[13] Wigge P, Kim M, Jaeger K, Busch W, Schmid M, Lohmann J, et al. Integration of spatial and temporal information during floral induction in Arabidopsis. Science. 2005; 309:1056-1059

[14] Yeung K, Seitz T, Li S, Janoschk P, McFerrank B, Kaiser C, et al. Suppression of Raf-1 kinase activity and MAP kinase signaling by RKIP. Nature. 1999;401:173-177

[15] Endo T, Shimada T, Fujii H, Kobayashi Y, Araki T, Omura M. Ectopic expression of an FT homolog from citrus confers an early flowering phenotype on trifoliate orange (*Poncirus trifoliata* L. Raf.). Transgenic Research. 2005;14:703-712

[16] Teper-Bamnolker P, Samach A. The flowering integrator FT regulates SEPALLATA3 and FRUITFULL accumulation in Arabidopsis leaves. Plant Cell. 2005;17:2661-2675

[17] Böhlenius H, Eriksson S, Parcy F, Nilsson O. (Retraction) The mRNA of the Arabidopsis gene FT moves from leaf to shoot apex and induces flowering. Science. 2007;316:367

[18] Tsuji H, Tachibana C, Tamaki S, Taoka K, Kyozuka J, Shimamoto K. Hd3a promotes lateral branching in rice. The Plant Journal. 2015;82:256-266

[19] Hayama R, Izawa T, Shimamoto K. Isolation of rice genes possibly involved in the photoperiodic control of flowering by a fluorescent differential display method. Plant and Cell Physiology. 2002;43:494-504

[20] Doi K, Izawa T, Fuse T, Yamanouchi U, Kubo T, Shimatani Z, et al. Ehd1, a B-type response regulator in rice, confers short-day promotion of flowering and controls FT-like gene expression independently of Hd1l. Genes & Development. 2004;18:926-936

[21] Xue W, Xing Y, Weng X, Zhao Y, Tang W, Wang L, Zhou H, Yu S, Xu C, Li X, Zhang Q. Natural variation in Ghd7 is an important regulator of heading date and yield potential in rice. Nature Genetics. 2008;40:761-767

[22] Izawa T. Daylength measurements by rice plants in photoperiodic short-day flowering. International Review of Cytology. 2007;256:191-222

[23] Matsubara K, Yamanouchi U, Wang Z, Minobe Y, Izawa T, Yano M. *Ehd2*, a rice ortholog of the maize *INDETERMINATE1* gene, promotes flowering by up-regulating *Ehd1*. Plant Physiology. 2008;148:1425-1435

[24] Park S, Kim S, Lee S, Je B, Piao H, Park S, et al. Rice *Indeterminate 1* (*OsId1*) is necessary for the expression of *Ehd1* (*Early heading date 1*) regardless of photoperiod. The Plant Journal. 2008;56:1018-1029

[25] Wu C, You C, Li C, Long T, Chen G, Byrne M, Zhang Q. RID1, encoding a Cys2/His2-type zinc finger transcription factor, acts as a master switch from vegetative to floral development in rice. Proceedings of the National Academy of Sciences of the United States of America. 2008;105:12915-12920

[26] Izawa T, Oikawa T, Tokutomi S, Okuno K, Shimamoto K. Phytochromes confer the photoperiodic control of flowering in rice (a short-day plant). The Plant Journal. 2000;22: 391-399

[27] Takano M, Inagaki N, Xie X, Yuzurihara N, Hihara F, Ishizuka T, Yano M, Nishimura M, Miyao A, Hirochika H, Shinomura T. Distinct and cooperative functions of phytochromes A, B, and C in the control of deetiolation and flowering in rice. Plant Cell. 2005;17:3311-3325

[28] Hoshikawa K. The Growing Rice Plant: An Anatomical Monograph. Japan: Nobunkyo; 1989. p. 225

[29] Taoka K, Ohki I, Tsuji H, Furuita K, Hayashi K, Yanase T, Yamaguchi M, Nakashima C, Purwestri YA, Tamaki S, Ogaki Y, Shimada C, Nakagawa A, Kojima C, Shimamoto K. 14-3-3 proteins act as intracellular receptors for rice Hd3a florigen. Nature. 2011;476:332-335

[30] Rosenquist M, Alsterfjord M, Larsson C, Sommarin M. Data mining the Arabidopsis genome reveals fifteen 14-3-3 genes. Expression is demonstrated for two out of five novel genes. Plant Physiology. 2001;127:142-149

[31] Chen F, Li Q, Sun L, He Z. The Rice 14-3-3 gene family and its involvement in responses to biotic and abiotic stress. DNA Research. 2006;13:53-63

[32] Yao Y, Du Y, Jiang L, Liu J. Molecular analysis and expression pattern of the 14-3-3 gene family. Journal of Biochemistry and Molecular Biology. 2007;40:349-357

[33] Purwestri YA, Ogaki Y, Tamaki S, Tsuji H, Shimamoto K. The 14-3-3 protein GF14c acts as a negative regulator of flowering in rice by interacting with the florigen Hd3a. Plant and Cell Physiology. 2009;50:429-438

[34] Muslin A, Xing H. 14-3-3 proteins: Regulation of subcellular localization by molecular interference. Cellular Signalling. 2000;12:703-709

[35] Igarashi D, Ishida S, Fukazawa J, Takahashi Y. 14-3-3 proteins regulate intracellular localization of the bZIP transcriptional activator RSG. Plant Cell. 2001;13(11):2483-2497

[36] Ishida S, Fukazawa J, Yuasa T, Takahashi Y. Involvement of 14-3-3 signaling protein binding in the functional regulation of the transcriptional activator REPRESSION OF SHOOT GROWTH by gibberellins. Plant Cell. 2004;16(10):2641-2651

[37] Lough T, Lucas W. Integrative plant biology: Role of phloem long-distance macromolecular trafficking. Annual Review of Plant Biology. 2006;57:203-232

[38] Giakountis A, Coupland G. Phloem transport of flowering signals. Current Opinion in Plant Biology. 2008;11:687-694

[39] Vert G, Nemhauser JL, Geldner N, Hong F, Chory J. Molecular mechanisms of steroid hormone signaling in plants. Annual Review of Cell and Developmental Biology. 2005;21:177-201

[40] Riechmann JL, Heard J, Martin G, Reuber L, Jiang C, Keddie J, Adam L, Pineda O, Ratcliffe OJ, Samaha RR, Creelman R, Pilgrim M, Broun P, Zhang JZ, Ghandehari D, Sherman BK, Yu G. *Arabidopsis* transcription factors: Genome-wide comparative analysis among eukaryotes. Science. 2000;290:2105-2110

[41] Kerstetter RA, Bollman K, Taylor A, Bomblies K, Poethig RS. *KANADI* regulates organ polarity in Arabidopsis. Nature. 2001;411:706-709

[42] Emery JF, Floyd SK, Alvarez J, Eshed Y, Hawker NP, Izhaki A, Baum SF, Bowman JL. Radial patterning of Arabidopsis shoots by class III HD-ZIP and KANADI genes. Current Biology. 2003;13:1768-1774

[43] Eshed Y, Izhaki A, Baum SF, Floyd SK, Bowman JL. Asymmetric leaf development and blade expansion in Arabidopsis are mediated by KANADI and YABBY activities. Development. 2004;131:2997-3006

[44] Purwestri YA, Ogaki Y Tsuji H, Shimamoto K. Functional analysis of OsKANADI1, a florigen Hd3a interacting protein in rice (*Oryza sativa* L.). Indonesian Journal of Biotechnology. 2012;17(2):169-176

[45] Abelenda JA, Navarro C, Prat S. From the model to the crop: Genes contolling tuber formation in potato. Current Opinion in Biotechnology. 2011;22:287-292

[46] Wigge PA. FT, a mobile developmental signal in plants. Current Biology. 2011;21: R374-R378

[47] Abelenda JA, Navarro C, Prat S. Flowering and tuberization: A tale of two nightshades. Trends in Plants Sciences. 2014;19:115-122

[48] Teo CJ, Takahashi K, Shimizu K, Shimamoto K, Taoka KI. Potato tuber induction is regulated by interactions between components of a tuberigen complex. Plant and Cell Physiology. 2017 February 1;58(2):365-374. DOI: 10.1093/pcp/pcw197

[49] Navarro C, Abelenda JA, Cruz-Oro E, Cuellar CA, Tamaki S, Silva J, Shimamoto K, Prat S. Control of flowering and storage organ formation in potato by FLOWERING LOCUS T. Nature. 2011;478:119-123

[50] Lee R, Samantha B, Fernand K, John M, Richard M. FLOWERING LOCUS T genes control onion bulb formation and flowering. Nature Communication. 2013;4:2884. DOI: 10.1038/ncomms3884

Chapter 5

ALSV Vector Substantially Shortens Generation Time of Horticultural Plants

Ichiro Kasajima, Chunjiang Li, Noriko Yamagishi and Nobuyuki Yoshikawa

Additional information is available at the end of the chapter

http://dx.doi.org/10.5772/intechopen.70317

Abstract

Flowering of plants is tightly regulated by both plant maturity and seasons in the year. Now that the *Flowering Locus T* (*FT*) gene has been revealed to encode the flowering hormone florigen, researchers are seeking to regulate and modify flowering behaviours by using florigen as a genetic tool. In place of transgenic approaches, *Apple latent spherical virus* (ALSV) vector was successful in promoting flowering of both model plants (Arabidopsis and tobacco), and fruit trees (e.g. apple, pear, and loquat), vegetables (e.g. tomato and cucumber), legumes (e.g. soybean), and ornamental flowers (e.g. petunia, Japanese gentian and Eustoma). In so doing, *FT* was expressed and/or *TFL1* was suppressed by the ALSV vector. ALSV is a latent (non-pathogenic) virus isolated from an apple tree. After induction of flowering and seed production in crops, ALSV is not transferred to most of the next-generation seedlings, or it can be artificially removed from the infected plant by incubation at high temperature. Thus, the generation times of horticultural plants are approximately halved, and the generation time of apple plants is substantially shortened to within one year. Hence, ALSV technology is expected to be useful as a part of New Plant Breeding Techniques (NPBT) for agricultural application.

Keywords: early flowering, florigen, FT, generation time, virus vector

1. Introduction

Breeding of horticultural plants can take several years or more. Plant cultivars are most frequently generated by crossing between different cultivars to combine various advantageous traits together, such as fruit/flower quality, pest tolerance and vigorous growth habit. Cultivars are often crossed with each other repeatedly to generate and fix such favourable traits. One generation time (from germination, flowering, seed set and germination of the next-generation seedling) is usually several months in herbaceous plants and several years in trees under field conditions [1]. Thus, crossing of herbaceous vegetables and ornamental flowers can be performed every year, while breeding of fruit trees may only be performed once every several years. If flowering could be accelerated in horticultural plants, we would be able to cross vegetables and flowers many times each year and cross fruit trees every year.

In order to cross vegetables several times in a year, problems related to the seasonality and day-length sensitivity of plant flowering also need to be addressed. Plants are either short-day, long-day or day-neutral (an aspect of the photoperiodism, the response to day length). Short-day plants flower in autumn and long-day plants flower in spring. Day-neutral plants flower in any growing season. Day length is now controlled by artificial lighting, but biotechnology to accelerate flowering will also solve the problem of day-length sensitivity without the need for regulating day length using lighting equipment. Alternatively, day-length sensitivity is also one of the important traits of horticultural plants. Ornamental flowers are harvested only when the plants set flowers. Fruits can be harvested only after the flowering seasons. Because of stable responses of horticultural plants to day length, farmers make great efforts to control time of flowering. Such artificial controls are possible only with lighting equipment, but the majority of crops are produced using sunlight alone. Gene modification by biotechnology such that a series of cultivars with different day-length responses are prepared for major horticultural cultivars will thus benefit agricultural production.

Basic studies using model plants have revealed the genes controlling flowering time in higher plants. After a functional gene is isolated in model plants, that gene and its homologs are often found to exert the same function in other plant species. This means that the flowering of vegetables, ornamental flowers and fruit trees can be controlled by this gene [2]. The name of the gene controlling plant flowering is '*Flowering locus T*' (*FT*). Initially, plant biologists aimed to express the *FT* gene in crops by transformation, which actually accelerated flowering [3–7]. This suggested the functional use of *FT* in controlling flowering time, but transgenic plants are not typically applicable to crop production. Plant biologists then attempted non-transgenic expression of *FT* in crops. To our knowledge, there have been no scientific papers demonstrating the control of flowering time by application of FT protein, *FT* DNA or *FT* RNA without genetic transformations or spontaneous mutations in plant genomic DNA. We used a virus vector derived from *Apple latent spherical virus* (ALSV) to deriver *FT* RNA and express FT protein in plant tissues. This system turned out to be very successful in promoting flowering in horticultural plants. This chapter explains in detail the characteristics of the ALSV vector, activities of *FT* and its related genes, and how the ALSV vector is used to promote flowering.

2. Characteristics of ALSV vector

ALSV is a rare, naturally occurring plant virus isolated from an apple tree. ALSV has only been detected in a single apple tree, excluding experimentally infected plants. The infected tree was grown in the orchard of a Japanese fruit tree research institute (Morioka, Iwate, Japan). The cultivar of this apple is 'Indo', named after Indiana State in the United States of America. The Indo apple used to be popular in Japan, but has now been largely replaced with other cultivars such as Fuji and Orin. ALSV does not naturally spread to nearby apple trees in the orchard [8], but it can infect the seedlings of apple cultivars in the experiments [9–11]. Moreover, ALSV is a 'latent' virus. This means that apple plants (and fruits) infected by ALSV do not show any discernible changes, such as yellow spots, inferior growths or twisted leaves. Thus, ALSV does not induce disease, but such latent characteristics are advantageous for plant biotechnology. With the ALSV vector, we can specifically upregulate (activate) or downregulate (inactivate) functions of target genes, for example, among the 57,000 genes in the apple genome [12], with the ALSV vector. There remain other advantages of the ALSV vector. ALSV evenly infects plant tissues and then enables even silencing (inactivation) of apple genes [11, 13]. ALSV can also infect other plants and upregulate/downregulate specific genes. For instance, ALSV can infect to soybean, petunia, pea, cucumber, pear, rose, Eustoma and Japanese gentian. ALSV is latent in many plant species [14, 15]. Unlike genetic transformation, whose infection rates greatly differ between cultivars of the same crop, ALSV does not typically select cultivars. Exceptions are tomato and almond, where ALSV can infect only part of the cultivars.

The protocol for preparation and infection of the ALSV vector is precisely described in our recent publication [15]. Briefly, plasmid DNAs for expression of ALSV RNAs are prepared by ordinary procedures, introduced into *Agrobacterium*, and 'Agro-inoculated' into benthamiana tobacco (*Nicotiana benthamiana*), which is readily infected by plant viruses. Virus particles are extracted from infected leaves; viral RNAs are extracted and introduced into horticultural plants by 'particle bombardment'. The precise structures of ASLV and ALSV vector are explained below.

ALSV consists of two genomic RNAs: ALSV-RNA1 and ALSV-RNA2. **Figure 1** shows the whole sequences of ALSV-RNA1 and ALSV-RNA2. The lengths of ALSV-RNA1 and ALSV-RNA2 are 6812 bases and 3384 bases, respectively. Similar to genomic RNAs of other plant viruses [16], genomic RNAs of ALSV encode single polyproteins. Polyproteins are translated as fused proteins and are digested into individual units by protease. ALSV-RNA1 encodes enzymes such as protease, helicase and RNA polymerase. ALSV-RNA2 encodes a movement protein and capsid proteins. The cleavage sites of the polyprotein were confirmed for ALSV-RNA2 by peptide sequences, whereas the cleavage sites of the polyprotein encoded by ALSV-RNA1 were just deduced from the peptide sequence [17]. The movement protein and the three capsid proteins are all necessary for cell-to-cell movement of ALSV within plant tissues [18].

ALSV is expected to be formed in plants such as *N. benthamiana* if ALSV-RNA1 and ALSV-RNA2 sequences are expressed simultaneously. Expression should not necessarily be performed by genetic transformation: faster and easier protocols for transient expression are usually adopted. Once ALSV is formed in the cells of *N. benthamiana*, it will regenerate and spread autonomously. Three sets of different plasmid vectors have been developed for ALSV vectors (**Table 1**).

Figure 1. Nucleotide sequences of ALSV genomic RNAs. Protein units are shown with different colours, between the non-coding sequences at the heads and the tails of ALSV genomic RNAs. The following proteins are encoded in ALSV-RNA1 in this order: PRO-co (protease cofactor), HEL (NTP-binding helicase), Vpg (viral protein genome-linked), C-PRO (cysteine protease), and POL (RNA polymerase). The following proteins are encoded in ALSV-RNA2 in this order: MP (movement protein), capsid proteins: Vp25, Vp20, and Vp24. Genbank/INSDC accessions and nucleotide lengths are shown in parentheses.

Vector	Expressed gene	Backbone	Antibiotics	Reference
pEALSR1	ALSV-RNA1	pE18PGT	Ampicillin	[9]
pEALSR2	ALSV-RNA2	pE18PGT	Ampicillin	
pBICAL1	ALSV-RNA1	pBICP35	Kanamycin	[19]
pBICAL2	ALSV-RNA2	pBICP35	Kanamycin	
pCALSR1	ALSV-RNA1	pCAMBIA1300	Kanamycin	[20]
pCALSR2	ALSV-RNA2	pCAMBIA1300	Kanamycin	
pBIN61:HC-Pro	HC-Pro (PVY)	pBIN61	Kanamycin	
pBE2113-HCPro	HC-Pro (ClYVV)	pBE2113	Kanamycin	[21]

From left column to right column are shown: vector names, expressed genes, plasmid backbone origins, antibiotics for plasmid selections, and reference papers.
PVY, *Potato virus Y*; ClYVV, *Clover yellow vein virus*.

Table 1. List of plasmid vectors.

pEALSR1 and pEALSR2 were the original vectors used for ALSV preparation. High concentrations of these vectors are prepared from colon bacteria (*Escherichia coli*) culture and rub-inoculated (inoculated with SiC, carborundum) onto the leaves of quinoa (*Chenopodium quinoa*) plants. The so-called agro-inoculation method was more convenient, and then new vectors pBICAL1 and pBICAL2 were developed, which are typically used in our experiments. These vectors are prepared in *E. coli* and transformed into Agrobacterium (*Agrobacterium tumefaciens*). *A. tumefaciens* harbouring ALSV vectors are injected into the leaves of *N. benthamiana* for transient expression of ALSV-RNAs. pCALSR1 and pCALSR2 undergo the same procedures with pBICAL1 and pBICAL2, but appear to merit higher concentrations in *E. coli*. The infection rate of 'wild-type' ALSV without modification is almost 100% in *N. benthamiana*. ALSV vectors with exogenous insertion sequences in either cloning sites suffer from much lower infection rates in *N. benthamiana*. To compensate for the reduced infection rate, 'silencing suppressor' protein HC-Pro derived from other viruses (ClYVV or PVY, [22]) is also transiently expressed when ALSV-RNAs are agro-inoculated into *N. benthamiana*. In conclusion, triple agro-inoculation of a mixture of ALSV-RNA1 clone, ALSV-RNA2 clone and HC-Pro clone enables efficient formation of ALSV vectors.

It was not easy to insert exogenous nucleotide sequences into ALSV. After many trials (Li C et al., unpublished data), three cloning sites were successfully used for nucleotide insertions (**Figure 2**) [9, 20]. One cloning site is located immediately after the stop codon of the polyprotein encoded by ALSV-RNA1 (named the SM site, after restriction sites for *Sal*I and *Mlu*I; **Figure 2A**). Another cloning site is located at the middle of the polyprotein encoded by ALSV-RNA2, between MP and Vp25 (named the XSB site, after restriction sites for *Xho*I, *Sma*I and *Bam*HI; **Figure 2B**). Nucleotide sequences around the XSB site are complicated. First, nucleotide sequences around the protease-digested site (QG site) encoding 10 amino acid residues (LLEGQGPDFT) were duplicated. Second, sequences of the restriction sites were introduced

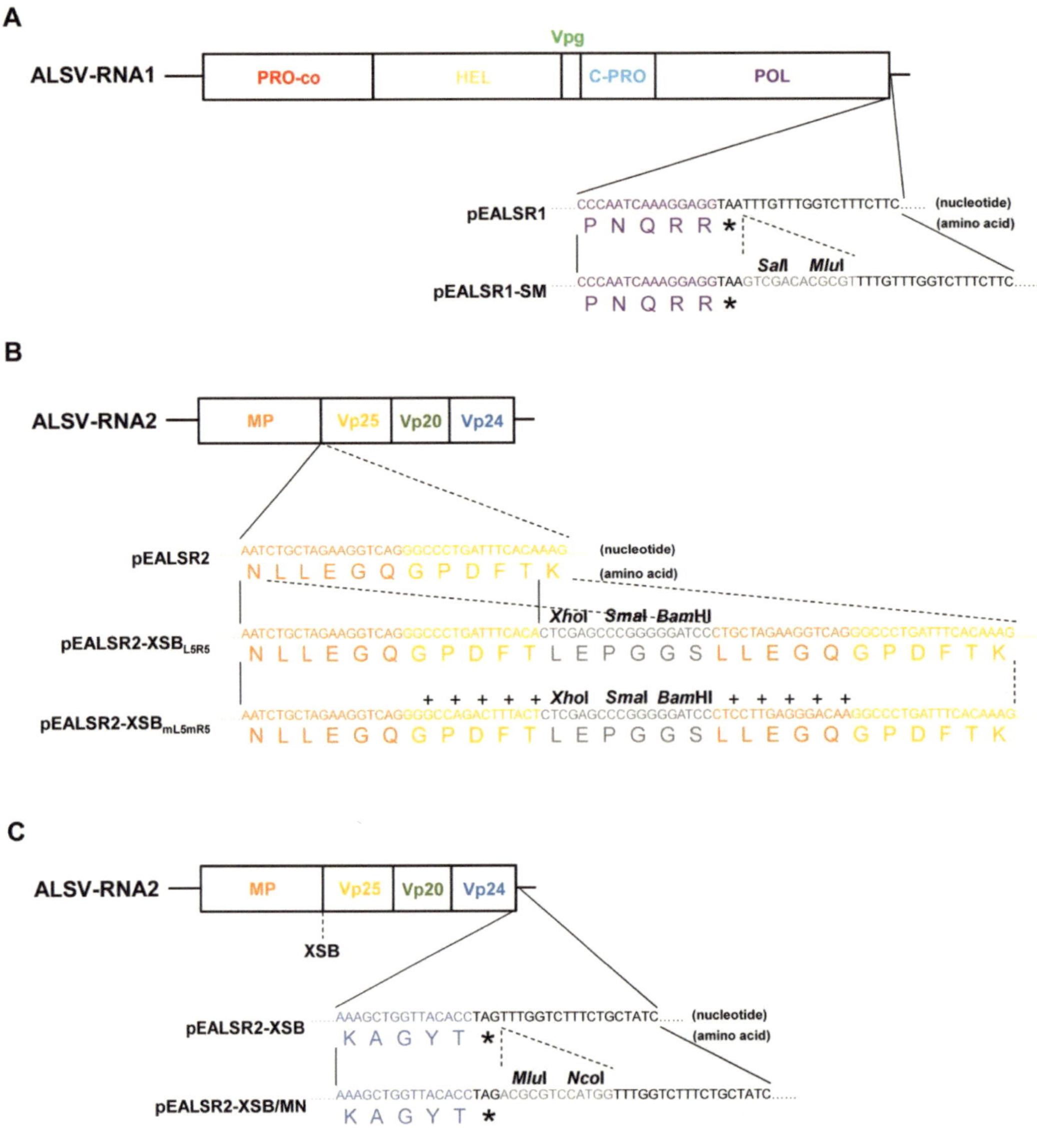

Figure 2. Structures of three cloning sites in ALSV vector. (A) The SM site near the tail of ALSV-RNA1. (B) The XSB site at the middle of the ALSV-RNA2. Introduced mutations are indicated with '+' letters. (C) The MN site near the tail of ALSV-RNA2. Introduced restriction sites are shown with grey letters in (A)–(C).

between the duplicated sequences. Finally, silent mutations (without amino acid changes) were introduced at the third nucleotides in each 10 codons surrounding the cloning site. These nucleotide modifications around the XSB site were necessary to insert exogenous nucleotide sequences at this site. The protease-digested site was duplicated such that both sites (before and after the inserted peptide) are cleaved by protease. Another special requirement for the virus vector is that the insertions are stably held by the virus. Nucleotide sequences in viruses, particularly repetitive sequences, are quite easily deleted (lost) probably via homologous recombination

[23–25]. To mitigate such unfavourable reactions, mutations had been introduced around the XSB site without changing the protease-digested amino acid sequence. The third cloning site is located immediately after the stop codon of the polyprotein encoded by ALSV-RNA2 (named the MN site, after the restriction sites of *Mlu*I and *Nco*I; **Figure 2C**). These cloning sites were constructed by Li C, based on pEALSR vectors, and then copied to the other vectors.

Among the three cloning sites of the ALSV vector (the SM site, the XSB site and the MN site), genes can be expressed only by their insertion into the XSB site. At the XSB site, the inserted genetic sequence is translated as a part of polyprotein, followed by digestion with protease. Cleaved proteins are expected to be attached with short peptides at both the N-terminus and the C-terminus , which derive from the protease-digested sites and the cloning site. These small attached peptides do not seem to affect the activity of inserted protein in most cases, as is experienced for transgenes introduced into transgenic plants.

All three cloning sites can be used for VIGS (virus-induced gene silencing). VIGS is a viral counterpart for gene silencing driven by antisense or inverted-repeat sequences expressed by genetic transformation [26, 27]. The virus itself is a natural target of gene silencing, then a nucleotide sequence inserted at any of the cloning sites of ALSV are silenced (degraded) by silencing mechanisms equipped in plant cells. Endogenous mRNAs harbouring the same nucleotide sequences with the inserted sequence in ALSV vector are also degraded, resulting in gene silencing. Any nucleotide sequences can be inserted at the SM site or at the MN site in principle. The nucleotide sequences inserted at the XSB site must be 'in frame' with the polyprotein, and this means the length of the inserts is a multiple of 3 and that each 'codon' encodes an amino acid without any stop codons (TAA, TAG or TGA as DNA sequences). Sizes of the inserted sequences are determined by the balance between the expected degree of silencing and potential risk of deletions (of the inserts from ALSV). Longer inserts will cause stronger silencing, but they will be more easily deleted from the ALSV vector. A length of 200 bases (or 201 bases at the XSB site) appears to be near the optimal size, with strong silencing and a relatively low possibility of deletions. Inserts shorter than 200 bases may not cause strong silencing, but the actual degree of silencing and frequency of deletions vary, depending on the inserted nucleotide sequences [14]. Insertion of nucleotide sequences at the XSB site is empirically simpler than insertions at the SM site or at the MN site. Insertions at the MN site strongly reduce the rate of viral infection, and they are also easily deleted. Insertions at the SM site have somewhat milder effects on infection rates and deletions. Insertions at the XSB site do not strongly reduce infection rate, but strongly suppress gene expression.

Model studies of silencing an endogenous gene with viral vectors often target the *Phytoene desaturase* (*PDS*) gene. **Figure 3** shows an *N. benthamiana* plant whose *PDS* gene was silenced by an ALSV vector. PDS catalyses the synthesis of carotenoids. Plants silenced in *PDS* expression become white, because they cannot accumulate both carotenoids and chlorophylls. This phenomenon (photo-bleaching) is believed to be caused by rapid 'photo-oxidation' of chlorophylls in the absence of photo-protective carotenoids in plant cells [28]. The representative mechanism of photo-protection by carotenoid would be non-photochemical quenching (NPQ) catalysed by PsbS protein and xanthophyll cycle. As Arabidopsis and rice mutants deficient in NPQ do not develop such white leaves caused by *PDS* silencing [29, 30], another major and unidentified photo-protective mechanism will be exerted by carotenoids in plant cells.

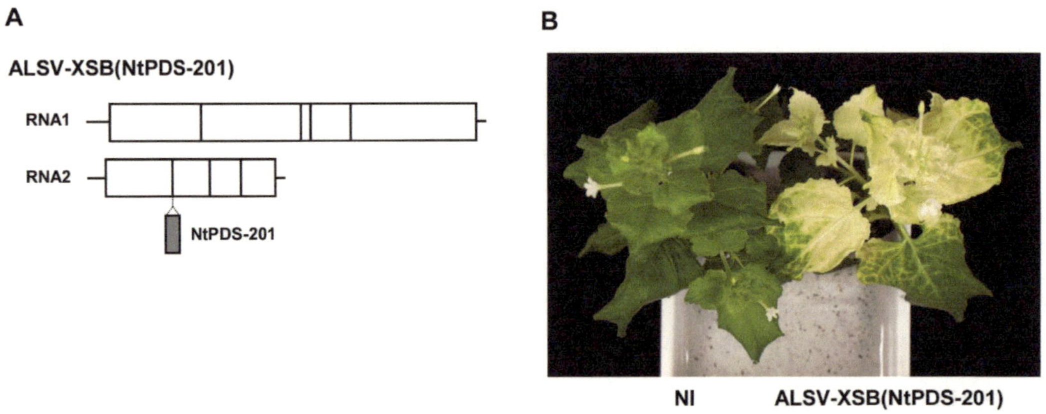

Figure 3. Silencing of *PDS* gene by ALSV vector. (A) An ALSV vector having insertion of a 201-base tobacco *PDS* gene fragment [ALSV-XSB(NtPDS-201)]. (B) The ALSV vector was infected into *N. benthamiana*. NI, non-inoculated ('healthy') plant.

Based on the degree of silencing of the apple *RubisCO small subunit* (*rbcS*) gene with the same 201-base insertion sequence, silencing effects of the three cloning sites differ from one another, and they are greater in the following order: the XSB site > the MN site > the SM site [31]. Again, insertion of a 201-base fragment of the target gene in the XSB site of the ALSV vector is recommended for efficient gene silencing with a lower risk of deletions. The SM site and the MN site are also available to silence additional target genes, by inserting different sequences at each cloning sites. Such simultaneous expression/suppression of different genes is possible, but such vectors typically show low infection rates. The MN site was also used for virus-induced transcriptional gene silencing (VITGS) in *N. benthamiana* and petunia, where upstream sequences (promoters) of the target genes are methylated by virus vectors [20]. The nucleotide sequence of an apple gene promoter was also mutated and inserted at the XSB site for infection to apple plants [15].

3. Early flowering of horticultural plants through expression of the *AtFT* gene

FT protein is a mobile signal of flowering. For example, FT induces flowering when a transgenic scion expressing the *FT* gene is grafted onto an *ft* mutant lacking *FT* gene function [32]. FT proteins produced in leaves and stems can contribute to early flowering, as well as FT proteins produced at shoot apices where flowers are formed. FT protein is originally produced in companion cells of the leaf vasculature. Then, the ALSV vector was prepared so that it expresses Arabidopsis *FT* gene (*AtFT*) at the XSB site (**Figure 4A**).

Horticultural plants, such as soybean and Eustoma infected by the ALSV vector expressing *AtFT* generate flowers earlier than normal (**Figure 4B** and **C**) [33, 34]. Such early-flowering plants often set seeds that germinate normally to generate next-generation seedlings. The ALSV vector has therefore successfully shortened the generation times of horticultural plants. Additional example of early-flowering plants is shown in **Figure 5**. These plants flowered at juvenile phase, showing the successful induction of early flowering by the ALSV-XSB(AtFT) vector.

Figure 4. Early flowering of soybean and Eustoma. (A) Schematic representation of the empty vector (wtALSV) and the vector expressing *AtFT* [ALSV-XSB(AtFT)]. (B) Soybean plants (cultivar 'Tanba-Guro') 1 month after inoculation of germinated seeds. (C) Eustoma (Lisianthus) plants 2 months after inoculation of 1-month-old seedlings.

Figure 5. Additional example of early-flowering plants. Petunia seedlings which exhibit early flowering via infection by the ALSV-XSB(AtFT) vector. Left plants are infected by wtALSV. The photograph was originally reported in Ref. [10].

4. Differential activities of *FT* genes in the ALSV vector

FT genes derived from various plant species are expressed by ALSV vectors in Arabidopsis, tobacco and soybean (**Figure 6A**) [11]. These *FT* genes are designated *AtFT* and *AtTSF* (derived from Arabidopsis), *VvFT* (grapevine), *GtFT1* and *GtFT2* (Japanese gentian), *InFT1* and *InFT2* (morning glory), *CsFT* (cucumber), *PmFT* (Japanese apricot), *GmFT2a* and *GmFT5a* (soybean), *LeFT* (tomato), *CiFT* (unshu mikan/satsuma mandarin), *PsFTa1* and *PsFTc* (pea), *MdFT1* and *MdFT2* (apple), *ComFT2* (squash) and *PtFT1* (aspen). Sizes of these FT proteins range from 172 to 184 amino acid residues, but their sizes have no correlation with their activities. A previous report described the degrees of inductions of Arabidopsis and tobacco flowering by these ALSV vectors [11]. **Figure 6B** summarizes these data, together with unpublished data on the induction of soybean flowering.

The average number of leaves at flowering was 30 in plants infected by the control wtALSV vector. The number of leaves was more than halved (less than 15) when plants were infected by the ALSV vectors expressing 10 *FT* genes, whereas, the number of leaves was more than 15 when plants were infected by the ALSV vectors expressing the remaining 9 *FT* genes. Among the 19 *FT* genes examined, *AtFT* and *VvFT* most strongly induced flowering, whereas *GtFT2* and *MdFT1* scarcely induced flowering. What determines the differences in the degree of flowering induction by these *FT* genes? **Figure 6C** shows the phylogenetic tree of the 19 FT proteins. The five FT proteins most strongly inducing flowering (indicated by circles) and the five FT proteins most weakly inducing flowering (indicated by triangles) are evenly distributed within the tree. This suggests that the activities of FT proteins were modified after species differentiation, rather than that their activities were evolutionarily fixed before species

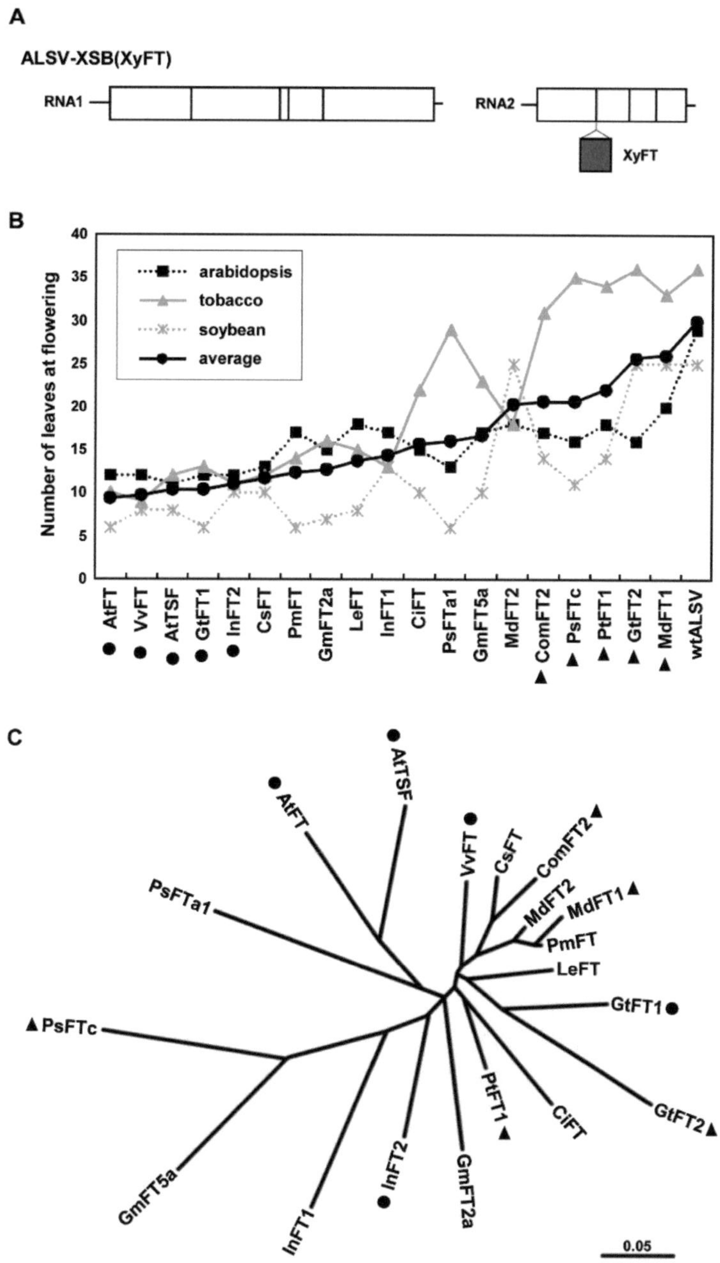

Figure 6. Induction of early flowering with various *FT* genes. (A) ALSV vectors used in this analysis. 'XyFT' represents *FT* genes derived from various plant species (such as *AtFT* and *MdFT1*). Respective *FT* genes are inserted at the XSB site in independent vectors. (B) Flowering times of ALSV-infected plants (Arabidopsis, tobacco or soybean) counted by the number of true leaves at flowering. Arabidopsis was grown at 25°C under a short-day condition (8 h:16 h light/dark photoperiod), and tobacco and soybean were grown at 25°C under a long-day condition (16 h:8 h light/dark photoperiod). 'Average' represents the average values of the numbers of leaves at flowering in Arabidopsis, tobacco and soybean. 'wtALSV' represents the wild-type ALSV vector without any insertion. MdFT2, GtFT2, MdFT1 and wtALSV did not induce soybean flowering, then the numbers of leaves are conveniently set as 25. (C) Phylogenetic tree of the FT proteins expressed by the ALSV vectors in this analysis. In (B) and (C), the five FT proteins most strongly inducing flowering are indicated by filled circles, and the five FT proteins most weakly inducing flowering are indicated by filled triangles.

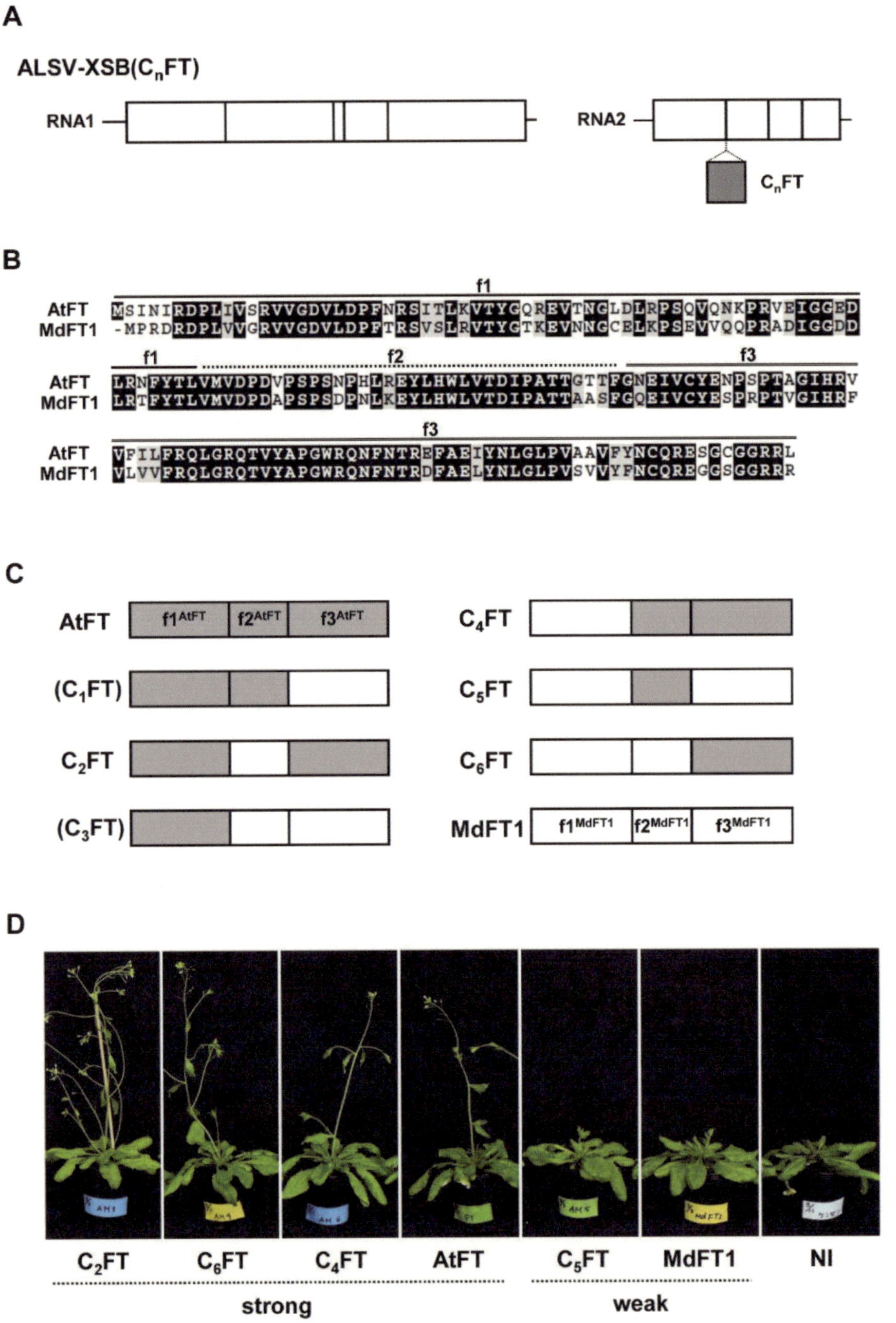

Figure 7. Activities of chimeric *FT* genes in Arabidopsis. (A) ALSV vectors used in this analysis. C_nFT, chimeric *FT* genes. (B) Amino acid sequences of AtFT and MdFT1, and the three fragments (f1, f2 and f3) defined in this analysis. (C) Combinations of FT fragments in chimeric FT proteins. The fragments derived from AtFT are grey, and the fragments derived from MdFT1 are white. (D) Arabidopsis plants infected by ALSV vectors. NI, non-inoculated plant. Juvenile plants with three true leaves were inoculated by particle bombardment of viral RNA. Viral infection was confirmed by RT-PCR (reverse transcription-polymerase chain reaction) analysis. Photographs were taken 30 days after inoculation.

differentiation. Even the *MdFT1* gene, the weakest *FT* gene in this analysis, more or less shortened flowering time when expressed in transgenic Arabidopsis and apple [35], confirming its activity as florigen. The activities of *FT* genes examined in our analysis are inconsistent with grafting experiments using transgenic Arabidopsis plants ectopically expressing *AtFT* or *AtTSF* [36]. In this paper, they found that AtTSF is scarcely mobile from rootstock to scion, then fails to induce early flowering. In contrast, AtTSF was highly active when expressed by the ALSV vector in **Figure 6**. The activity of FT proteins as the regulator of gene expressions may matter when expressed by ALSV vectors, rather than their mobility along the phloem.

Chimeric *FT* genes between *AtFT* (strongest FT) and *MdFT1* (weakest FT) were constructed and investigated in order to determine which part of the *FT* gene regulates its activity. This analysis was performed by Yamagishi N, and is presented for the first time here. The ALSV vectors in this analysis express chimeric *FT* genes at the middle of the polyprotein in RNA2 (**Figure 7A**). **Figure 7B** shows the amino acid sequences of AtFT and MdFT1. FT proteins were divided into three fragments (f1, f2 and f3); f1 corresponds to the first exon, f2 corresponds to the second and the third exons and f3 corresponds to the fourth exon of the FT genes. Chimeric FT proteins are designated C_1FT, C_2FT, C_3FT, C_4FT, C_5FT and C_6FT (**Figure 7C**). All ALSV vectors including those expressing AtFT and MdFT1 were inoculated into *N. benthamiana*, but the vectors expressing C_1FT and C_3FT did not infect *N. benthamiana*. When virus-infected Arabidopsis plants were grown under a long-day condition, C_2FT, C_6FT, C_4FT and AtFT strongly induced flowering, while C_5FT and MdFT1 only weakly induced flowering (**Figure 7D**). This shows that the C-terminus (f3) of the FT protein determines its strength as a florigen. Although we typically use the ALSV vector expressing *AtFT* to shorten generation time in horticultural plants, chimeric *FT* genes such as *C_2FT* many more strongly induce flowering in some plant species.

5. FT/TFL1 gene family

FT is a member of the FT/TFL1 gene family (also called PEBP, Phosphatidylethanolamine-binding protein). This gene family consists of five subgroups in plants: FT (Flowering locus T), TFL1 (Terminal flower 1), CEN (Centroradialis), BFT (Brother of FT) and MFT (Mother of FT). All Arabidopsis genes (*AtFT, AtTFL1, AtATC, AtBFT* and *AtMFT*) and apple genes belonging to FT/TFL1 family are displayed in a phylogenetic tree of protein sequences, together with some representative genes from other plant species (**Figure 8**). Names of apple genes and their Genbank/the International Nucleotide Sequence Database Collaboration (INSDC) accessions are as follows: MdFT1 (BAD08340.1), MdFT2 (NP_001280810.1), MdTFL1-1 (NP_001280887.1), MdTFL1-2 (NP_001280794.1), MdCENa (NP_001280813.1), MdCENb (NP_001280940.1), MdBFTa (XP_008376539.1), MdBFTb (NP_001280770.1), MdMFTa (XP_008374830.1) and MdMFTb (NP_001281044.1).

Among the five subgroups of FT/TFL1 gene family, FT and TFL1 regulates initiation (induction) of floral organ development and then regulates the time of flowering. FT protein is produced before the time of flowering and carried to shoot apices. As already described, FT positively regulates (increases) the expression of genes such as *SOC1* (*SUPPRESSOR OF OVEREXPRESSION OF CO 1*) and *AP1* (*APETALA1*) to induce flowering. In contrast, TFL1 is expressed at shoot apices and negatively regulates (decreases) gene expression to suppress

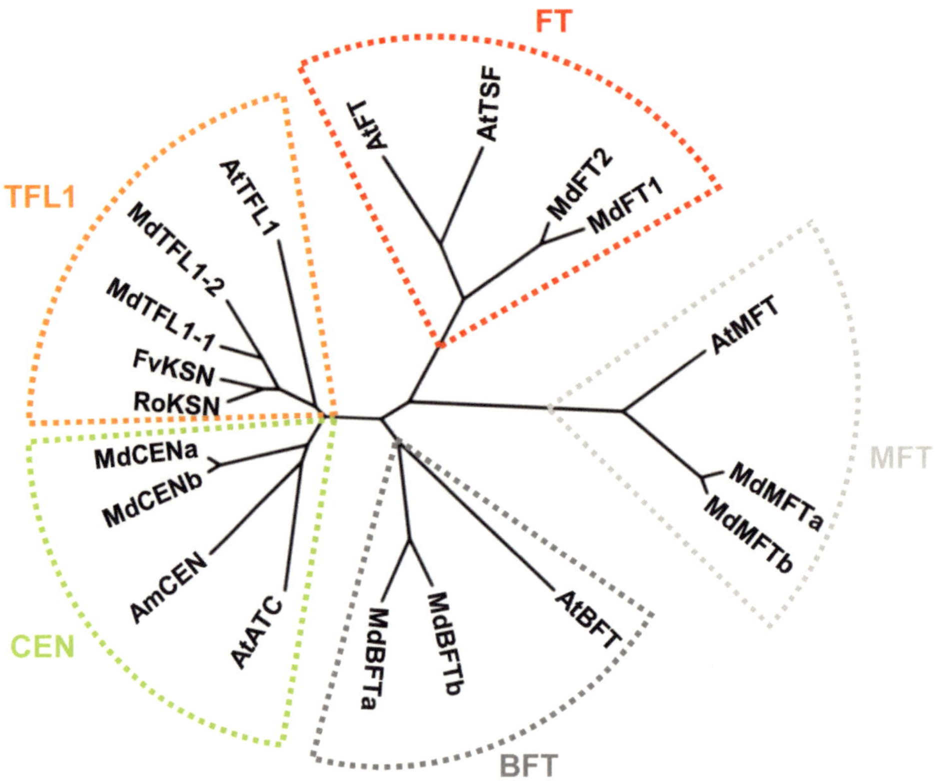

Figure 8. Phylogenetic tree of FT/TFL1 family genes. Five subfamilies are indicated by sectors.

flowering [37]. Thus, the balance between FT and TFL1 expressions have important roles in determining the time of flowering, and determining the parts of plant shoots where the flowers are generated. For example, the mutant of strawberry *FvKSN* gene and the mutant of rose *RoKSN* gene (the TFL1 family genes) can both flower in any growing season (continuous flowering), whereas ordinary strawberry and rose plants generate flowers only at one specific season of the year (seasonal flowering) [38]. The subgroup CEN is represented by the snapdragon *AmCEN* gene. This subgroup does not regulate flowering time, but does regulate the architecture of inflorescence (flower clusters at shoot apices). Snapdragon usually generates 'indeterminate' inflorescence, where many flowers are repeatedly generated on the side of the inflorescence, and the inflorescence continues to elongate without generating a flower at the very apex of the inflorescence. On the other hand, a snapdragon mutant in the *AmCEN* gene has 'determinate' inflorescence, where a limited number of flowers are generated on the side of the inflorescence, and the inflorescence stops elongating with a flower generated at the very apex of the inflorescence (called terminal flower). Thus, CEN negatively regulates flowering at the apex of the inflorescence to generate indeterminate inflorescence [39].

As apparent in the phylogenetic tree (**Figure 8**), apple has two copies of each five FT/TFL1 subgroup genes. The functions of these apple genes are not completely clear, but there are several reports on their activities and expression patterns. *MdFT1* and *MdFT2* are both expressed in apple plants. *MdFT1* is suggested to regulate the flowering time of apple plants [35], but direct evidence has not yet been found. Expression levels of *MdCENa* are much greater than those of *MdCENb* [40]. Considering its classification into the CEN subgroup and high expression levels, *MdCENa* may regulate the inflorescence architecture of apple plants. Thus, apple plants generate clustered flowers at the apices of newly developing branches. The size of the flower clusters may be smaller in the apple mutants of the *MdCENa* gene. *MdTFL1-1* and *MdTFL1-2* share similar expression patterns. Their expression is reduced when floral buds are initiated in July [40]. According to the observations explained in the next section, *MdTFL1-1* regulates the flowering time of apple plants.

6. Early flowering by combination between *FT* expression and *TFL1* suppression in apple

As already described, early flowering of horticultural plants can be achieved by expression of highly active *FT* genes such as *AtFT* with ALSV vectors (**Figures 4–6**). AtFT induces the early flowering of apple plants at a rate of 30% when expressed by the ALSV vector, but MdFT1 does not induce early flowering in any apple plant [10]. This result reconfirms that highly active FT homologs should be expressed by ALSV vectors to achieve early flowering, rather than endogenous but weakly active FT homologs. Consistent with this idea, other highly active FT homologs (AtTSF and GtFT1) also induced early flowering in apple plants [11]. Unlike herbaceous plants such as tobacco and soybean, the early-flowering rate of apple plants was as low as 30%, even with the most highly active *FT* genes. This difference between plant species may be related to their natural intervals from germination to flowering (months in herbs and years in trees).

Early flowering of plants may be also achieved through suppression of the *TFL1* gene, the negative regulator of flowering. *MdTFL1-1* was silenced in apple plants with ALSV vector [13]. Early flowering was observed, but in only 10% of the infected apple plants. To further improve the early-flowering rate of apple plants, simultaneous expression of *AtFT* and suppression of *MdTFL1* homologs were performed [11]. The early-flowering rate was not increased when *MdTFL1-2* was suppressed simultaneously with the expression of *AtFT*, but the early-flowering rate increased to 90% when *MdTFL1-1* was suppressed simultaneously with the expression of *AtFT*. Thus, early flowering of apple plants was successful at high rates with the combination between *AtFT* expression and *MdTFL1-1* suppression. As well as early flowering at high rates, part of the early-flowering apple plants obtained by *AtFT* expression/*MdTFL1-1* suppression continuously generates flowers on branches, whereas the early-flowering apple plants obtained by only *AtFT* expression generate flowers only once. This difference in flowering traits may be also caused by negative regulation of flowering by *MdTFL1-1* in young apple plants. An example of early-flowering apple plants obtained by simultaneous *AtFT* expression and *MdTFL1-1* suppression is shown in **Figure 9**. Early-flowering apple plants set fruits and seeds after pollination with the pollen gathered from other compatible cultivars. Apple fruits typically mature at about 6 months after flowering. The next-generation seedlings germinated within 1 year, counting from the germination of the mother plant [11].

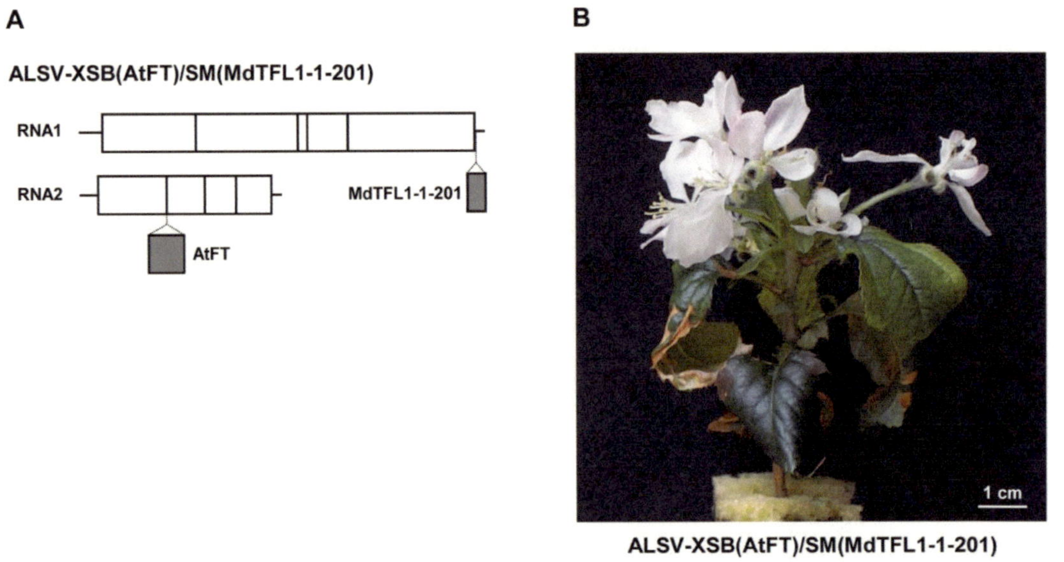

Figure 9. Early-flowering apple plant. (A) ALSV vector used in this analysis. 'MdTFL1-1-201' represents a 201-base fragment of the *MdTFL1-1* gene. (B) Early-flowering apple plant (seedling of Orin progeny: 'progeny' means 'next-generation'). Photograph was taken 67 days after inoculation of viral RNA to germinated seed.

7. Conclusion

ALSV will be among the most useful viral vectors for genetic engineering of horticultural plants, although ALSV is not presently applicable to cereals. Like other viral vectors, inserted sequences into the ALSV cloning sites are easily deleted upon infection to plants, depending on the sizes and sequences of the inserted sequences. What is more, ALSV vectors with insertions sometimes do not infect the plants. However, these problems can be managed by technical efforts and the selection of the insertion sequences. Nucleotide fragments larger than 1 kb can even be introduced into the XSB site of the ALSV vector and infected to plants. We hope that this chapter furthered general understanding of the structure and function of the ALSV vector, and promote its use in both basic and applied studies.

Early flowering through infection of ALSV vectors shortens the generation times of horticultural plants. This technique is expected to promote breeding of horticultural plants. It may not be a popular concept, but there seems to be strongly active *FT* genes (such as *AtFT*) and weakly active *FT* genes (such as *MdFT1*) according to our experiments. The analysis of chimeric *FT* genes in ALSV vector indicated that the difference in the activities of *AtFT* and *MdFT1* is determined by the C terminus, rather than the N terminus. Both types of *FT* genes will have ecological advantages in specific plants species, but highly active types of FTs are useful for early flowering of horticultural plants with ALSV vectors. In addition, simultaneous expression of *FT* and suppression of *TFL1* is beneficial for high rate of early flowering and continuous flowering. ALSV is not transferred to most of the next-generation plants, so the next-generation plants are neither transgenic nor infected by ALSV. Therefore, ALSV can be used as a new plant breeding technique (NPBT).

Acknowledgements

We like to acknowledge Kiyoaki Maeda, Tomiko Kawaguchi, all laboratory members and collaborators for helpful supports in virus infection to horticultural plants.

Funding

This study was supported by the Science and Technology Research Promotion Program for Agriculture, Forestry, Fisheries and Food Industry (Japan).

Author details

Ichiro Kasajima, Chunjiang Li, Noriko Yamagishi and Nobuyuki Yoshikawa*

*Address all correspondence to: yoshikawa@iwate-u.ac.jp

Faculty of Agriculture, Iwate University, Iwate, Japan

References

[1] van Nocker S, Gardiner SE. Breeding better cultivars, faster: Applications of new technologies for the rapid deployment of superior horticultural tree crops. Horticulture Research. 2014;1:14022

[2] Blümel M, Dally N, Jung C. Flowering time regulation in crops—What did we learn from Arabidopsis? Current Opinion in Biotechnology. 2015;32:121-129

[3] Li WM, Tao Y, Yao YX, Hao YJ, You CX. Ectopic over-expression of two apple *Flowering Locus T* homologues, *MdFT1* and *MdFT1*, reduces juvenile phase in *Arabidopsis*. Biologia Plantarum. 2010;54:639-646

[4] Tanaka N, Ureshino A, Shigeta N, Mimida N, Komori S, Takahashi S, Tanaka-Moriya Y, Wada M. Overexpression of *Arabidopsis FT* gene in apple leads to perpetual flowering. Plant Biotechnology. 2014;31:11-20

[5] Xing W, Wang Z, Wang X, Bao M, Ning G. Over-expression of an *FT* homolog from *Prunus mume* reduces juvenile phase and induces early flowering in rugosa rose. Scientia Horticulturae. 2014;172:68-72

[6] Li C, Zhang Y, Zhang K, Guo D, Cui B, Wang X, Huang X. Promoting flowering, lateral shoot outgrowth, leaf development, and flower abscission in tobacco plants overexpressing cotton *FLOWERING LOCUS T* (*FT*)-like gene *GhFT1*. Frontiers in Plant Science. 2015;6:454

[7] Klocko AL, Ma C, Robertson S, Esfandiari E, Nilsson O, Strauss SH. *FT* overexpression induces precocious flowering and normal reproductive development in *Eucalyptus*. Plant Biotechnology Journal. 2016;14:808-819

[8] Nakamura K, Yamagishi N, Isogai M, Komori S, Ito T, Yoshikawa N. Seed and pollen transmission of Apple latent spherical virus in apple. Journal of General Plant Pathology. 2011;77:48-53

[9] Li C, Sasaki N, Isogai M, Yoshikawa N. Stable expression of foreign proteins in herbaceous and apple plants using Apple latent spherical virus RNA2 vectors. Archives of Virology. 2004;149:1541-1558

[10] Yamagishi N, Sasaki S, Yamagata K, Komori S, Nagase M, Wada M, Yamamoto T, Yoshikawa N. Promotion of flowering and reduction of a generation time in apple seedlings by ectopical expression of the *Arabidopsis thaliana FT* gene using the Apple latent spherical virus vector. Plant Molecular Biology. 2011;75:193-204

[11] Yamagishi N, Kishigami R, Yoshikawa N. Reduced generation time of apple seedlings to within a year by means of a plant virus vector: A new plant-breeding technique with no transmission of genetic modification to the next generation. Plant Biotechnology Journal. 2014;12:60-68

[12] Velasco R, et al. The genome of the domesticated apple (*Malus* x *domestica* Borkh.). Nature Genetics. 2010;42:833-839

[13] Sasaki S, Yamagishi N, Yoshikawa N. Efficient virus-induced gene silencing in apple, pear and Japanese pear using *Apple latent spherical virus* vectors. Plant Methods. 2011;7:15

[14] Igarashi A, et al. *Apple latent spherical virus* vectors for reliable and effective virus-induced gene silencing among a broad range of plants including tobacco, tomato, *Arabidopsis thaliana*, cucurbits, and legumes. Virology. 2009;386:407-416

[15] Kasajima I, Ito M, Yamagishi N, Yoshikawa N. *Apple latent spherical virus* (ALSV) vector as a tool for reverse genetic studies and non-transgenic breeding of a variety of crops. In: Rajewsky N, Jurga S, editors. Plant Epigenetics, RNA Technologies Series. Springer International Publishing AG, Cham, Switzerland; 2017. pp. 513-536

[16] Wellink J, van Kammen A. Proteases involved in the processing of viral polyproteins. Archives of Virology. 1988;98:1-26

[17] Li C, Yoshikawa N, Takahashi T, Ito T, Yoshida K, Koganezawa H. Nucleotide sequence and genome organization of Apple latent spherical virus: A new virus classified into the family *Comoviridae*. Journal of General Virology. 2000;81:541-547

[18] Yoshikawa N, Okada K, Asamuma K, Watanabe K, Igarasi A, Li C, Isogai M. A movement protein and three capsid proteins are all necessary for the cell-to-cell movement of apple latent spherical cheravirus. Archives of Virology. 2006;151:837-848

[19] Kawai T, Gonoi A, Nitta M, Kaido M, Yamagishi N, Yoshikawa N, Tao R. Virus-induced gene silencing in apricot (*Prunus armeniaca* L.) and Japanese apricot (*P. mume* Siebold

& Zucc.) with the Apple latent spherical virus vector system. Journal of the Japanese Society for Horticultural Science. 2014;83:23-31

[20] Kon T, Yoshikawa N. Induction and maintenance of DNA methylation in plant promoter sequences by apple latent spherical virus-induced transcriptional gene silencing. Frontiers in Microbiology. 2014;5:595

[21] Yaegashi H, Yamatsuta T, Takahashi T, Li C, Isogai M, Kobori T, Ohki S, Yoshikawa N. Characterization of virus-induced gene silencing in tobacco plants infected with apple latent spherical virus. Archives of Virology. 2007;152:1839-1849

[22] Csorba T, Kontra L, Burgyán J. Viral silencing suppressors: Tools forged to fine-tune host-pathogen coexistence. Virology. 2015;479-480:85-103

[23] Nagy PD, Bujarski JJ. Efficient system of homologous RNA recombination in brome mosai virus: Sequence and structure requirements and accuracy of crossovers. Journal of Virology. 1995;69:131-140

[24] Nagy PD, Simon AE. New insights into the mechanisms of RNA recombination. Virology. 1997;235:1-9

[25] Li C, Yoshikawa N. Virus-induced gene silencing of *N* gene in tobacco by *Apple latent spherical virus* vectors. In: Uyeda I, Masuta C, editors. Plant Virology Protocols: Methods in Molecular Biology. Vol. 1236. Humana Press, New York, USA; 2014. pp. 229-240

[26] Cogoni C, Macino G. Conservation of transgene-induced post-transcriptional gene silencing in plants and fungi. Trends in Plant Science. 1997;2:438-443

[27] Frizzi A, Huang S. Tapping RNA silencing pathways for plant biotechnology. Plant Biotechnology Journal. 2010;8:655-677

[28] Kumagai MH, Donson J, Della-Cioppa G, Harvey D, Hanley K, Grill LK. Cytoplasmic inhibition of carotenoid biosynthesis with virus-derived RNA. Proceedings of the National Academy of Sciences of the United States of America. 1995;92:1679-1683

[29] Havaux M, Niyogi K. The violaxanthin cycle protects plants from photooxidative damage by more than one mechanism. Proceedings of the National Academy of Sciences of the United States of America. 1999;96:8762-8767

[30] Kasajima I, Ebana K, Yamamoto T, Takahara K, Yano M, Kawai-Yamada M, Uchimiya H. Molecular distinction in genetic regulation of nonphotochemical quenching in rice. Proceedings of the National Academy of Sciences of the United States of America. 2011;108:13835-13840

[31] Yamagishi N, Li C, Yoshikawa N. Promotion of flowering by *Apple latent spherical virus* vector and virus elimination at high temperature allow accelerated breeding of apple and pear. Frontiers in Plant Science. 2016;7:171

[32] Notaguchi M, et al. Long-distance, graft-transmissible action of *Arabidopsis* FLOWERING LOCUS T protein to promote flowering. Plant & Cell Physiology. 2008;49:1645-1658

[33] Yamagishi N, Yoshikawa N. Expression of *FLOWERING LOCUS T* from *Arabidopsis thaliana* induced precocious flowering in soybean irrespective of maturity group and stem growth habit. Planta. 2011;233:561-568

[34] Fekih R, Yamagishi N, Yoshikawa N. Apple latent spherical virus vector-induced flowering for shortening the juvenile phase in Japanese gentian and lisianthus plants. Planta. 2016;244:203-214

[35] Kotoda N, et al. Molecular characterization of *FLOWERING LOCUS T*-like genes of apple (*Malus* x *domestica* Borkh.). Plant & Cell Physiology. 2010;51:561-575

[36] Jin S, Jung HS, Chung KS, Lee JH, Ahn JH. FLOWERING LOCUS T has higher protein mobility than TWIN SISTER OF FT. Journal of Experimental Botany. 2015;66:6109-6117

[37] Ratcliffe OJ, Amaya I, Vincent CA, Rothstein S, Carpenter R, Coen ES, Bradley DJ. A common mechanism controls the life cycle and architecture of plants. Development. 1998;125:1609-1615

[38] Iwata H, et al. The *TFL1* homologue *KSN* is a regulator of continuous flowering in rose and strawberry. The Plant Journal. 2012;69:116-125

[39] Bradley D, Carpenter R, Copsey L, Vincent C, Rothstein S, Coen E. Control of inflorescence architecture in *Antirrhinum*. Nature. 1996;379:791-797

[40] Mimida N, Kotoda N, Ueda T, Igarashi M, Hatsuyama Y, Iwanami H, Moriya S, Abe K. Four *TFL1/CEN*-like genes on distinct linkage groups show different expression patterns to regulate vegetative and reproductive development in apple (*Malus* x *domestica* Borkh.). Plant & Cell Physiology. 2009;50:394-412

Section 4

Improving the Structural Properties of the Plant

Chapter 6

Development of Transgenic Sorghum Plants with Improved *In Vitro* Kafirin Digestibility

Lev A. Elkonin, Julia V. Italyanskaya,
Valery M. Panin and Nikolai Yu. Selivanov

Additional information is available at the end of the chapter

http://dx.doi.org/10.5772/intechopen.69973

Abstract

Improvement of nutritional value of crops is one of the main goals of plant biotechnology. These studies are extremely important for sorghum—a unique drought-tolerant cereal crop that is of special importance for sustainable grain production in the arid regions. The major cause of relatively low nutritive value of sorghum grain is the resistance of one of its seed storage proteins, γ-kafirin, to protease digestion. Using *Agrobacterium*-mediated genetic transformation, we have obtained transgenic sorghum plants harboring a genetic construct for RNA interference (RNAi) silencing of the γ-kafirin gene. In T_1 generation, transgenic plants with modified endosperm texture were found. These plants had lowered level of the 28-kDa γ-kafirin protein and kafirin oligomers, which are formed by natural kafirin polymerization. *In vitro* protein digestibility analysis showed that the amount of undigested protein in transgenic plants was reduced by 2.9–3.2 times, in comparison with the original line, the digestibility index reached 85–88% (60% in the original line). HPLC analysis showed that total amino acid content in transgenic plants was reduced, while the lysine proportion was increased by 1.6–1.7 times. PCR analysis confirmed inheritance of the genetic construct up to T_4 generation.

Keywords: transgenic plants, *Agrobacterium*-mediated genetic transformation, gamma-kafirin, *in vitro* protein digestibility, RNA silencing, endosperm, *Sorghum bicolor* (L.) Moench

1. Introduction

Development of plant varieties and hybrids that possess the necessary traits and properties is the main goal of plant breeding. With the accumulation of knowledge in the field of genetics, physiology and molecular biology of plants, the ability of breeders and geneticists to create valuable varieties and hybrids has significantly expanded. Development of genetic engineering approaches have allowed creating a significant number of cultivars and lines, resistant to biotic and abiotic stresses, with improved quality of final products, increased photosynthetic rate and nutrient-use efficiency [1].

Creation of transgenic plants with the changed composition of proteins and improved nutritional value is one of the most promising areas of genetic engineering. These investigations are particularly relevant for cereals being the main source of food and feed protein. It is known that humans receive from cereals up to 50% of proteins (or up to 70% in developing countries) and up to 65% of calories, in which the storage proteins account for up to 80% of the total protein content in the mature seed [2]. To solve this problem, various genetic engineering technologies had been developed. These technologies allow the introduction of new genes and thereby modulate the synthesis of new proteins with higher nutritional value or in a highly specific way downregulate genes that control the synthesis of proteins with a low nutritional value or reducing the digestibility or assimilation of other proteins [3–5]. Genetic engineering techniques are quite promising for enrichment of cereal grain with essential amino acids, i.e., lysine, tryptophan and methionine [6]. To date, transgenic lines with a modified composition of seed storage proteins, with increased lysine content and with improved baking properties have already been obtained in all most important species of cereals—maize, rice and wheat [7–9].

These studies are extremely important for sorghum—a unique drought-tolerant cereal crop having special importance for sustainable grain production in the arid regions. Today, sorghum is one of the five most widely cultivated cereal crops, and with the increase of climate aridity, observed in many regions of the globe, demand for sorghum will increasingly grow. However, the majority of sorghum cultivars and hybrids have relatively poor nutritive value in comparison with other cereals [10, 11]. One of the reasons of relatively low nutritive value of sorghum grain is resistance of its seed storage proteins (kafirins) to protease digestion [12]. The causes of the poor sorghum protein digestibility were studied extensively [10, 13, 14]. Among the factors that cause or may affect this phenomenon, there are chemical structures of kafirin molecules, some of which (α- and β-kafirins) are abundant with sulfur-containing amino acids capable to form S–S bonds, resistant to protease digestion; interactions of kafirins with non-kafirin proteins and non-protein components such as polyphenols and polysaccharides; spatial organization of different kafirins in the protein bodies of endosperm cells; endosperm structure (vitreous or floury).

It is generally accepted that the peripheral disposition of γ-kafirin in protein bodies reduces digestibility of α-kafirin—the major sorghum seed storage protein located central position in protein bodies and comprising up to 80% of total endosperm kafirins [13, 14]. This hypothesis is supported by studies of protein bodies of the mutant with improved protein digestibility. In this mutant, protein bodies shape has been changed from spherical to invaginate; the γ-kafirin was located at the bottom of invaginations where it should not interfere with the digestion of the α-kafirin [15]. Recent study also showed that a sorghum mutant with high digestibility of kafirins has a point mutation in the signal sequence of the α-kafirin gene, which apparently disrupts its deposition in protein bodies [16]. One of the main characteristic features of kafirin proteins is their ability to form oligomers or polymers of high molecular weight. These oligomers comprise α- and γ-kafirins that are linked together by disulfide (S–S) bonds [17, 18]. They are resistant to protease digestion and occur more in the vitreous endosperm fraction [14, 19].

Improving of sorghum genetic transformation technology [20, 21] makes it possible to solve this problem by using RNA interference (RNAi) that allows targeted downregulation of individual genes. In recent years, RNAi technology has become widely used for changing the composition of the storage proteins and starch in different cereal species [3–5].

In maize, with using of genetic constructs harboring inverted repeats of genes of α-zeins (19 and 22 kDa), transgenic lines with suppressed synthesis of these proteins were obtained [22, 23]. It was found that repression of the synthesis of zeins possessing a relatively low nutritional value leads to accumulation of other proteins with a higher nutritional value. Maize plants with gene silencing of α-zeins were characterized by doubled content of essential amino acids tryptophan and lysine in the kernels. These experiments showed that gene silencing of 22 kDa α-zein resulted in the formation of the floury endosperm. Such a modification in the type of endosperm was associated with abnormalities in the formation of the structure of protein bodies, namely the violation of deposition of 19 kDa α-zein into the center of a protein body, or a modification of its interaction with β- and γ-zeins [22].

In sorghum, transgenic lines with genetic constructs capable of RNAi silencing of different kafirin classes were obtained [24–27]. Transgenic plants harboring these constructs were characterized by improved *in vitro* protein digestibility (IVPD) that was accompanied by opaque floury endosperm. Unfortunately, the floury endosperm reduces the practical value of these lines, because the reduction of the vitreous layer increases the fragility of kernels and increases the susceptibility to fungal infection.

In our experiments, we obtained transgenic sorghum plants with genetic construct for silencing of the gamma-kafirin gene [28]. These plants retained sectors of vitreous endosperm in their kernels and were characterized by high level of *in vitro* kafirins digestibility. In this chapter, we review these experiments and present new data, confirming inheritance of the genetic construct and its effect on endosperm protein spectrum and endosperm texture.

2. Obtaining of transgenic plants with genetic construct for RNA silencing of the γ-kafirin gene

To obtain transgenic plants with silencing of gamma-kafirin gene, the binary silencing vector, pNRKAFSIL, has been designed. This vector contained a hairpin insert that consisted of an inverted repeat of the fragment of the γ-kafirin gene and *ubi1* intron as the spacer between the arms of the inverted repeat (**Figure 1**). The 307-bp fragment of the γ-kafirin gene was

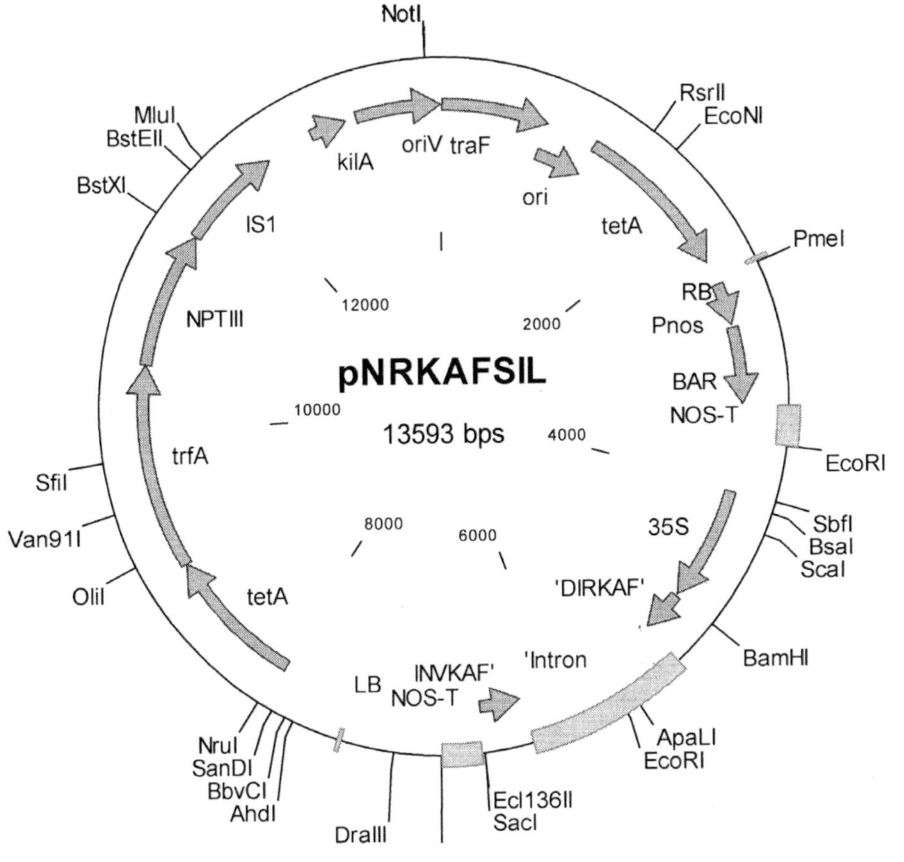

Figure 1. Map of the pNRKAFSIL vector containing hairpin insert consisted from inverted repeat of the fragment of the γ-kafirin gene ("INVKAF" and "DIRKAF") and *ubi1* intron as the spacer between the arms of the inverted repeat (published with the permission of the publishing house "Nauka").

isolated by PCR from genomic DNA of sorghum. The sequence corresponded to bases 280–588 of GeneBank accession number M73688 [29]. This construct was driven by the CaMV 35S-promoter. The T-DNA region of this vector contained selectable marker *bar* gene driven by *nos*-promoter. The binary vector pNRKAFSIL was introduced in *Agrobacterium tumefaciens* GV3101.

To obtain transgenic plants with genetic construct for RNA silencing of the γ-kafirin gene, cocultivation of immature embryos of sorghum cv. Zheltozernoe 10 (Zh10) (15–17 days after pollination) with cell suspension of the *A. tumefaciens* strain GV3101/pNRKAFSIL was performed.

Activation of *vir*-genes was made according to the published protocol [30] with some modifications. *A. tumefaciens* strain GV3101/pNRKAFSIL vector was grown on an Agrobacterium (AB) minimal medium [31] with the antibiotics for 3 days at 28°C. After that a loop of the Agrobacterium cells were transferred into the flask with 20 ml of liquid yeast extract peptone (YEP) medium with the antibiotics and grown for 9 h under continuous shaking (220 rpm) at

28°C. Then, the cells were collected by centrifugation and suspended in a small volume (5–6 ml) of modified AB medium without phosphates with the addition of 200 μM acetosyringone (Sigma-Aldrich, USA) and were incubated for 18 h under gentle shaking (60–70 rpm) at 22–23°C. After incubation, the cells were collected by centrifugation and suspended in inoculating medium PHI-I [32] with the addition of 200 μM acetosyringone to a final OD_{600}=0.6. This suspension was used for inoculation of immature embryos.

Agrobacterial transformation was based on previously published protocols [20, 32] with some modifications. Immature embryos after pre-cultivation for 3 days on the agar M11 medium [33] were placed onto sterile filter paper wetted with inoculating medium and were inoculated with an agrobacterial cell suspension in PHI-I medium for 10 min at room temperature. The Agrobacterium inoculum was then removed, and the filter with embryos was transferred into another Petri dish on a dry filter and was wetted with cocultivation medium (M11 medium supplemented with 200 μM acetosyringone). The cocultivation step was performed for 3 days at 23 ± 1°C in the dark. After cocultivation, the embryos were transferred to the M11 medium with the addition of 200 mg/l timentin solidified with 2.5 g/l phytagel and were cultured at 27 ± 1°C in the dark for 7 days. Then, the embryos with developing embryogenic calli were subcultured to the fresh medium of the same composition with the addition of 2.5 mg/l glufosinate ammonium (GA) and were cultivated at 28°C in the dark for 3–4 weeks.

From two experiments on cocultivation of immature sorghum embryos of Zh10 with *A. tumefaciens* strain GV3101/pNRKAFSIL 35 embryogenic calli survived after selection on the medium with 2.5 mg/l GA (**Table 1**; **Figure 2A**). For plant regeneration, the herbicide-tolerant calli were transferred onto regeneration medium (murashige and skoog (MS), 1.0 mg/l kinetin, 1.0 mg/l Indole-3-Acetic Acid (IAA)) and maintained at 25°C under a photoperiod of 16 h light and 8 h dark. Initiation of shoot development was observed in 13 calli transferred to the regeneration medium, but in the majority of the cultures, shoot development was arrested at early stages. Nevertheless, few regenerants were obtained (**Figure 2B**), one of which turned out to be PCR-positive in the experiment with primers to the *bar* gene (**Figure 3A**).

Experiment	Number of embryos	Number of EC resistant to 2.5 mg/l GA	Number of cultures with regenerants	Number of T_0 plants (PCR-positive)[1]	Number of plants in T_1 generation		Number of plants in T_2 generation[2]	
					Total	Resistant to 2.5 mg/l GA (PCR-positive)	Total	Resistant to 2.5 mg/l GA (PCR-positive)
#1	49	21	11	3 (1)	40	10 (6 out of 6 studied)	141	103 (10 out of 19 studied)
#2	31	14	2	1 (0)	–	–	–	–

Notes: EC = embryogenic cultures; GA = glufosinate ammonium.
[1]PCR with primers to *bar* gene.
[2]Combined progeny from PCR-positive plants from T_1 generation.

Table 1. Selection of transgenic plants by cocultivation of immature sorghum embryos of Zheltozernoe 10 with the *A. tumefaciens* GV3101/pNRKAFSIL.

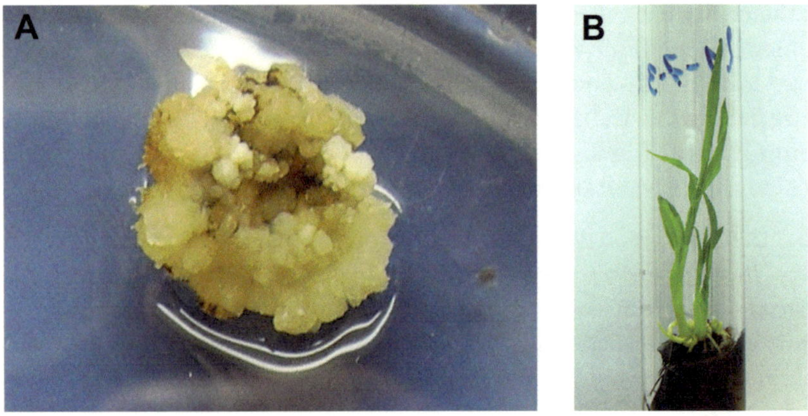

Figure 2. Embryogenic callus developing on M11 medium with 2.5 mg/l glufosinate ammonium (A) and regenerated plants (B) obtained in experiment on *Agrobacterium*-mediated genetic transformation of immature sorghum embryos with *A. tumefaciens* strain GV3101/pNRKAFSIL.

Self-pollinated progeny (T_1) of this plant (#94) was tested for herbicide tolerance by germination on a medium containing 2.5 mg/l of the selective agent (**Figure 4**). This concentration causes browning and death of sensitive non-transgenic plants. Herbicide-tolerant plants were found, and the sensitive plants predominated over tolerant ones (**Table 1**). Some of herbicide-tolerant plants that were tested with the primers to the *bar* gene were proved to be PCR-positive (**Figure 3B**). In the progeny of PCR-positive T_1 plants (i.e., in the T_2 generation) that were grown on the medium with 2.5 mg/l GA, the frequency of herbicide tolerant plants was significantly higher (**Table 1**) and some of these plants were also PCR-positive (data not shown).

These data testify that the progeny of plant #94 inherited the transgenic construct. A low frequency of tolerant plants in the T_1 generation might be explained by silencing of the *bar* gene driven by *nos*-promotor because silencing of transgene is a common phenomenon in sorghum

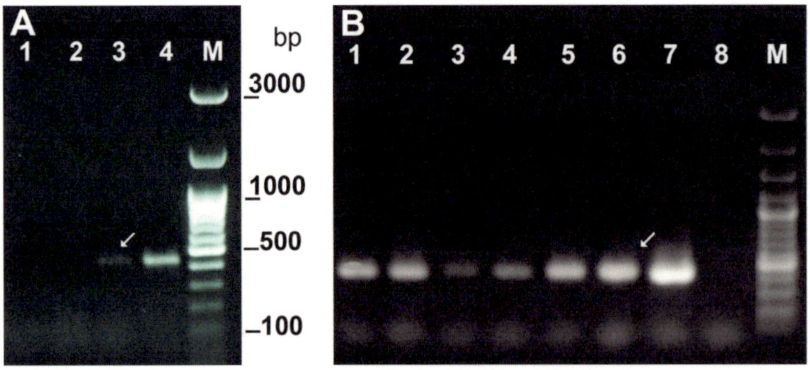

Figure 3. PCR analysis of genomic DNA of plants from T_0 (A) and T_1 (B) generations obtained by genetic transformation with *A. tumefaciens* GV3101/pNRKAFSIL with primers to *bar* gene. (A) 1—original non-transgenic line, Zheltozernoe 10; 2—negative control without template DNA; 3—T_0 plant (#94); 4—pNRKAFSIL; M—100-bp ladder. (B) 1–6—individual plants from T_1 generation; 7—pNRKAFSIL; 8—negative control without template DNA; M—100-bp ladder. Amplified fragment of the *bar* gene (444 bp) is marked by arrow (published with the permission of the publishing house "Nauka").

Figure 4. Segregation for tolerance to 2.5 mg/l glufosinate ammonium in the progeny of PCR-positive plant #94 obtained by genetic transformation with *A. tumefaciens* GV3101/pNRKAFSIL. Note green tolerant plants and necrosis in sensitive plants (tolerant plants survived selection, have been transferred from the agar medium to tap water to improve their survival in soil).

genetic transformation [34]. In the T_2 generation, segregation of GA-tolerant vs. GA-sensitive plants corresponds to a monogenic ratio 3:1 ($\chi^2 = 0.286$; $0.50 < P < 0.75$) (**Table 1**).

The inheritance of T-DNA in subsequent generations, including T_4, was confirmed by PCR analysis using primers to the marker gene *bar*, with each of the three T_2 families studied contained PCR-positive plants (**Figure 5A**).

To verify the presence of the genetic construct for RNA silencing of the γ-kafirin gene in the transgenic plants, we performed a PCR analysis of a number of plants from T_3 and T_4 generations for the presence of ubiquitin intron. In the studied plants, amplification of a fragment of this gene was observed, which confirmed the presence of a genetic construction for γ-kafirin silencing in the genome of the obtained transgenic plants (**Figure 5B**).

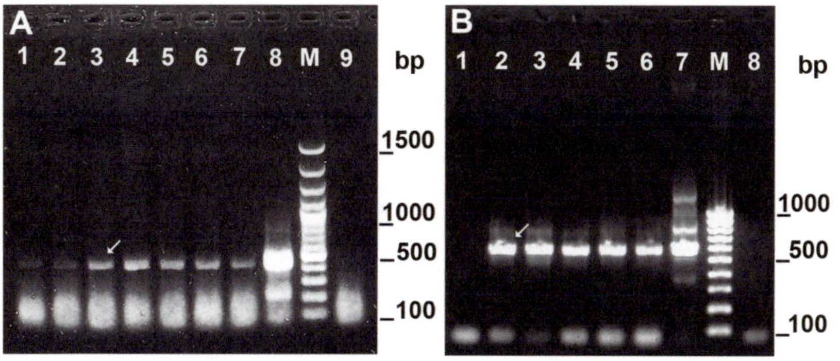

Figure 5. PCR analysis of genomic DNA of sorghum plants from T_3 and T_4 generations obtained by genetic transformation with *A. tumefaciens* GV3101/pNRKAFSIL with primers to *bar* gene (A) and *Ubi*-intron (B). (A) 1—T_3 94-2-04-1; 2—T_3 94-2-04-3; 3—T_4 94-2-11-2-4; 4—T_3 94-3-04-3; 5—T_4 94-3-08-2-1; 6—T_4 94-3-08-2-3; 7—T_4 94-2-11-2-1; 8—pNRKAFSIL; M—100-bp ladder and 9—negative control (without DNA template). Amplified fragment of the *bar* gene (444 bp) is marked by arrow. (B) 1—Zh10, original non-transgenic line; 2—T_3 94-2-04-1; 3—T_4 94-2-11-2-1; 4—T_0 Ogonek; 5—T_3 94-2-04-2; 6—T_4 94-3-08-3-3; 7—pNRKAFSIL; M—100-bp ladder and 8—negative control (without DNA template). Amplified fragment of *Ubi*-intron (584 bp) is marked by arrow.

3. Analysis of electrophoretic spectra of endosperm proteins in plants with genetic construct for RNA silencing of the γ-kafirin gene

To identify the expression of the introduced genetic construct, the experiments on SDS-PAGE of endosperm proteins were performed. The samples (20 mg of flour) were incubated with a sample buffer (0.0625 M Tris·HCl, pH 6.8) under reducing conditions (2% SDS, 5% β-mercaptoethanol, destroying the S–S bonds of kafirin polymers) or in native, non-reducing conditions (without β-mercaptoethanol) at 100°C for 90 s. The samples were centrifuged, and supernatant was used for SDS-PAGE in 13.0% (w/v) polyacrylamide gel (PAG) according to modified Laemmli method [35]. The gels were stained with Coomassie Brilliant Blue R-250. The electrophoretic spectra were carefully studied, and particular attention was paid to the γ-kafirin content, the suppression of which was to be expected, and to content of kafirin oligomers (≈47 and ≈66 kDa), which consist from α- and γ-kafirins [14, 17, 18].

It was found that in kernels of original non-transgenic line Zh10 content of polypeptides with Mr ≈47 and ≈66 kDa was markedly higher than in transgenic plants. These differences were observed both in SDS-PAGE performed in non-reducing conditions (**Figure 6**) and in reducing conditions (see Section 5).

Notably, electrophoresis in non-reducing conditions revealed that the level of polypeptide corresponding to γ-kafirin (28 kDa, marked by an arrow) in transgenic plants was significantly lowered compared to the original non-transgenic line, which was to be expected with the silencing of the γ-kafirin gene. In addition, as we found previously in experiments on SDS-PAGE in reducing conditions, content of α-kafirin monomers (25 and 23 kDa) was also

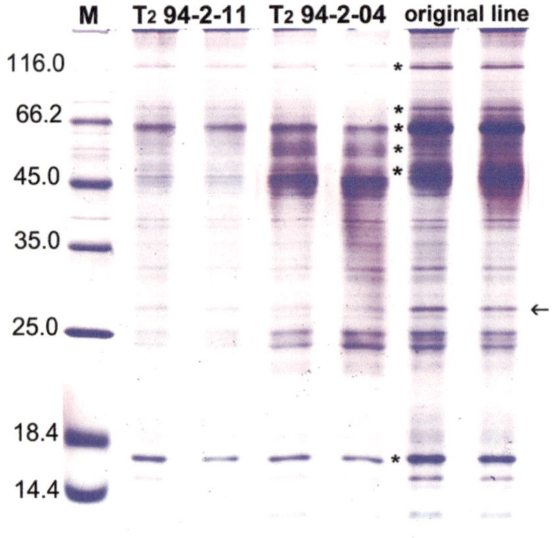

Figure 6. SDS-PAGE of endosperm proteins of transgenic sorghum plants with genetic construct for silencing of the γ-kafirin gene in non-reducing conditions. γ-kafirin is marked by arrow; the proteins, the amount of which varies in the original line and in transgenic plants, are marked by asterisks.

reduced in transgenic plants ([28], see Section 5). Perhaps the suppression of the synthesis of γ-kafirin caused also effect on synthesis or accumulation of α-kafirins.

Noteworthy, lowered amount of protein in the lanes of transgenic plants (**Figure 6**) is not an artifact, since in each sample the same amount of flour was taken in the study; all samples were subjected to the same treatment, and the same amount of extract was taken when carrying out SDS-PAGE. Therefore, such reduced protein content is due to the genetic characteristics of the samples. A similar decrease in protein content was observed in transgenic maize plants carrying constructs for RNA silencing of γ-zein [36].

4. Endosperm texture in plants with genetic construct for RNA silencing of the γ-kafirin gene

It is known that one of the consequences of silencing of γ-prolamins in maize and sorghum is a disruption of the formation of the vitreous layer of the endosperm. In previously obtained transgenic sorghum lines with genetic constructs for γ-kafirin silencing [24, 26, 27], as well as in the mutant with high digestibility [37], the kernels had a floury endosperm type. In transgenic maize plants, silencing of γ-zein also resulted in reduction of the vitreous layer and the formation of floury endosperm that suggests its role in interaction with starch granules and in the formation of the vitreous endosperm [36]. In this connection, we paid special attention to the endosperm texture in the kernels of our transgenic plants.

Careful examination of the kernels developed on panicles of T_1 plants obtained in our experiments revealed three plants, #94-3, #94-4 and #94-6, in which the kernels with almost floury endosperm were found (**Figure 7A**) [28]. Such kernels clearly differed from those of the original non-transgenic line, which have a thick vitreous layer (**Figure 7B**). The amount of such kernels varied in different panicles of one and the same T_1 plants. For example, in T_1 plant, #94-2, all kernels developed on its first panicle did not express floury phenotype, although kernels on its second panicle had either almost floury or modified structure of endosperm. In such kernels, the vitreous layer was significantly reduced and developed as sectors or blurs surrounded by floury endosperm (**Figure 7C–E**). Remarkably, these kernels resemble the kernels of recombinant sorghum lines obtained by hybridization of highly digestible mutant with floury endosperm (*hdhl*) with ordinary sorghum lines with low protein digestibility and vitreous endosperm [38]. Formation of this endosperm type in our transgenic plants apparently reflects peculiarities of expression of inserted genetic construct during kernel development.

Modified endosperm type of plant #94-2 inherited for three generations and was observed in $T_2$and T_3 families (94-2-04; 94-2-05 and 94-2-11) characterized by high *in vitro* protein digestibility (see Section 5), although kernels with thin or irregularly developed vitreous endosperm (**Figure 7F–H**) also formed in panicles of plants from these families. The plants from T_2 and T_3 families from the progeny #94-3 (94-3-04; 94-3-08) had both modified, irregularly developed and normal vitreous endosperm types.

No variation of endosperm type was observed in the kernels developed in other PCR-positive T_1 plants, #94-1 and #94-5.

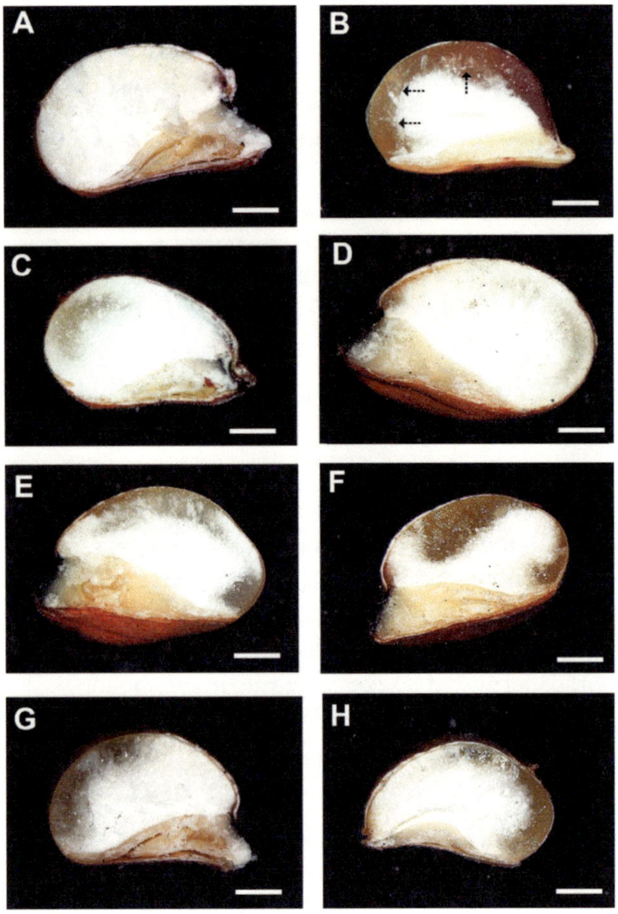

Figure 7. Cross sections of kernels with different types of endosperm of transgenic sorghum plants with genetic construct for silencing of the γ-kafirin gene. (A) Kernel with floury endosperm (T_3 94-2-05-1); (B) kernel of original non-transgenic line Zheltozernoe 10 with thick vitreous endosperm (marked by arrows); (C–E) modified endosperm type with blurs and sectors of vitreous endosperm (T_2 94-2-05, T_2 94-2-04, T_1 94-6, respectively); (F–H) irregularly developed vitreous endosperm (T_2 94-3-08; T_3 94-2-05-2; T_3 94-2-11-2, respectively). Bar = 1 mm.

5. *In vitro* digestibility of endosperm proteins

To study *in vitro* protein digestibility, the method of whole-grain flour pepsin treatment, widely practiced in the past few years, was used [37–42]. The flour (20 mg) of transgenic samples (kernels of transgenic plants from T_1–T_3 generations) and of original non-transgenic line Zh10 was treated with 5 ml of 0.15% pepsin solution (Sigma-Aldrich, activity: 806 units/mg of protein) in a 0.1 M potassium phosphate buffer (pH 2.0) for 120 min at 37°C with repeated shaking. The control samples were incubated in potassium phosphate buffer without pepsin addition under the same conditions. For quantitative estimation of protein digestibility, the digested and control samples were centrifuged and the pellet was incubated with a sample buffer (0.0625 M Tris·HCl, pH 6.8) under reducing conditions (see above). The samples were subjected to SDS-PAGE (see above). After electrophoresis, the gels were scanned. The amount of protein, expressed as volume (intensity × area) of kafirin bands or of total protein bands in the lane, was quantified with the Scangel program (Dr. A.F. Ravich, Agricultural Research Institute of the South-East Region, Saratov, Russian Federation) [41]. The digestibility value was counted as the percent ratio of the difference between protein volume in the control sample and in digested sample to the protein volume in the control sample. All experiments were performed in two replications.

It was found that transgenic plants obtained in our experiments significantly differed in digestibility of endosperm storage proteins from the original non-transgenic line Zh10 [28]. Comparison of electrophoretic spectra before and after pepsin digestion of proteins of T_1 plant #94-2 (almost floury endosperm; **Figure 8A**, lanes 1, 2) with Zh-10 kernels (**Figure 8A**, lanes 5, 6) revealed that in transgenic plant the amount of undigested α-kafirin monomers and total undigested protein was significantly fewer (in 1.7–1.9 times) than in original non-transgenic line (**Table 2**). The digestibility value reached 85.4%, whereas in original line this value was about 60%, usual index for sorghum flour (**Table 3**). Remarkably, in kernels of transgenic plant #94-3-08 (T_2 generation) with thick irregularly developed vitreous endosperm (**Figure 8A**, lanes 3, 4), the differences in kafirin digestion, in comparison with original line Zh-10 (**Figure 8A**, lanes 5, 6), were more pronounced: the amount of undigested monomers was 17.5 times fewer, and the amount of total undigested protein was 4.7 times fewer than in original line (**Table 2**). The digestibility value reached 92% (**Table 3**).

One should note considerable differences in content of kafirin oligomers between original non-transgenic line Zh10 and transgenic plants (**Figure 8**). Decreased content of kafirin oligomers, which apparently was caused by reduction of γ-kafirin synthesis, might be the reason of higher protein digestibility in transgenic plants.

Another examples of significantly improved kafirin digestibility in transgenic plants obtained in our experiments are presented in **Figure 8B**, where almost complete disappearance of kafirin monomers after pepsin digestion was observed in plants from T_2 generation with both floury (#94-2-11, lanes 5, 6) and modified endosperm (#94-2-04, lanes 1, 2, and #94-2-05, lanes 3, 4). Total protein digestibility indices in 94-2-05 and 94-2-11 plants reached 74.1% and 90.7%, respectively, that significantly differed from original non-transgenic line (**Table 3**). Remarkably, in electrophoretic spectra of digested samples of transgenic plants, one should note the polypeptides with molecular weights approx. 40 and 42 kDa. Previously, we found that these polypeptides were more prominent in electrophoretic spectra of more digestible lines than in spectra of poorly digestible ones [41]. In this study, appearance of these polypeptides in transgenic samples coincides with almost complete digestion of kafirin monomers and slightly reduces total protein digestibility values (**Table 3**).

Plants from T_3 generation inherited improved digestibility of kafirins. Comparison of electrophoretic spectra of proteins obtained from plants #94-2-11-2 and # 94-2-11-3 (**Figure 9A**, lanes 1–4), which were characterized by almost floury or modified endosperm, with the spectrum of the original line (**Figure 9A**, lanes 5, 6) before and after pepsin digestion showed that in transgenic plants, the amount of undigested α-kafirin monomers was significantly fewer (3.4–6.0 times, respectively) (**Table 2**). Likewise, the total sum of undigested proteins was also reduced (2.9–3.2 times). The digestibility value reached 85.5–87.8%, whereas in the original line this value was 59.3%, the usual index for sorghum flour (**Table 3**).

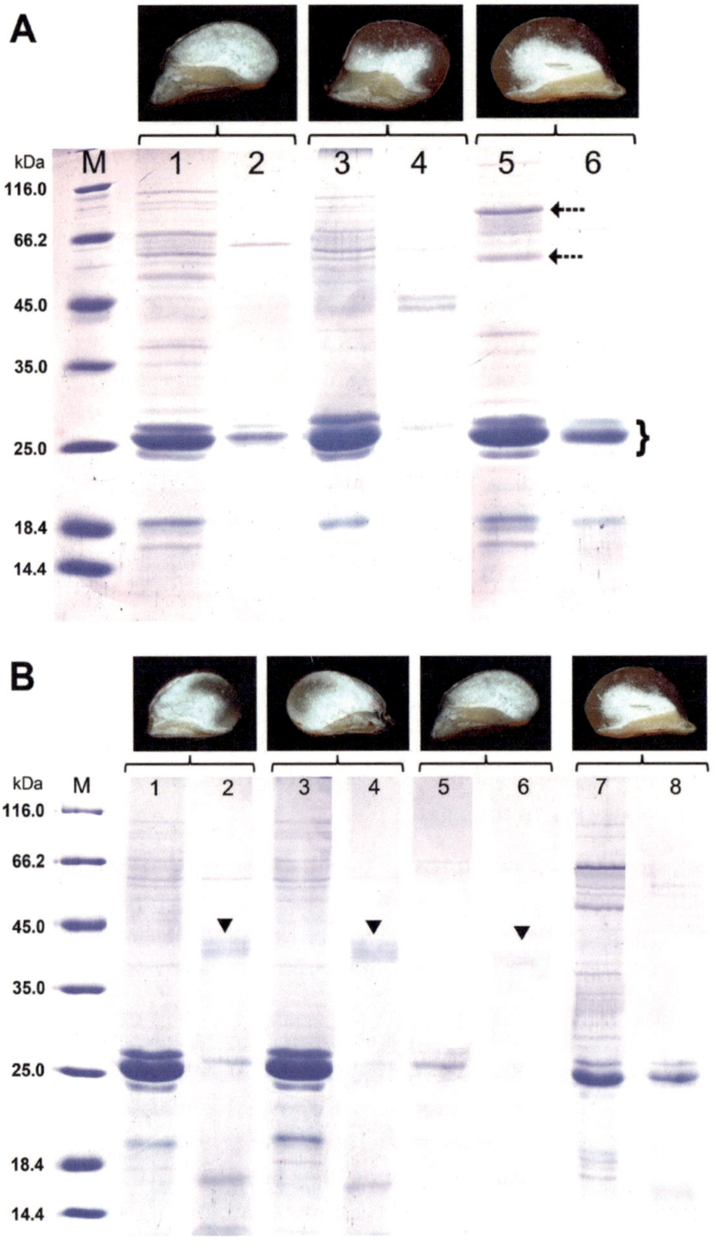

Figure 8. SDS-PAGE of endosperm proteins of kernels developed on transgenic sorghum plants with genetic construct for silencing of the γ-kafirin gene in reducing conditions. (A) 1, 2—#94-2 (T_1 generation) with almost floury endosperm; 3, 4—#94-3-8 (T_2 generation) with thick vitreous endosperm; 5, 6—original non-transgenic line Zheltozernoe 10 (Zh10) with normal vitreous endosperm; M—molecular weight markers (kDa; Thermo Scientific). 1, 3, 5—before, and 2, 4, 6—after pepsin digestion. Dashed arrows indicate probable kafirin oligomers. α-kafirin monomers are indicated by brace. (B) 1, 2—#94-2-04; 3, 4—#94-2-05, both with modified endosperm, in which vitreous layer is covered by thin floury layer (**Figure 4C**); 5, 6—#94-2-11 with floury endosperm; 7, 8—original non-transgenic line Zh10. 40 and 42 kDa appeared in digested samples are marked by arrows. 1, 3, 5, 7—before and 2, 4, 6, 8—after pepsin digestion (**Figure 8A** is published with the permission of the publishing house "Nauka").

Plant	Lane[1]	Estimated protein quantity[2]		Percent of undigested protein[3]	
		α-kafirin monomers	total	α-kafirin monomers	total
Figure 8A					
T_1 94-2	1 (c)	$4.887 \cdot 10^6$	$15.083 \cdot 10^6$	23.8	15.15
	2 (p)	$1.163 \cdot 10^6$	$2.285 \cdot 10^6$		
T_2 94-3-08	3 (c)	$8.166 \cdot 10^6$	$19.077 \cdot 10^6$	2.6	5.45
	4 (p)	$0.340 \cdot 10^6$	$1.925 \cdot 10^6$		
Zheltozernoe 10 (original line)	5 (c)	$5.124 \cdot 10^6$	$11.899 \cdot 10^6$	45.4	25.9
	6 (p)	$2.328 \cdot 10^6$	$3.079 \cdot 10^6$		
Figure 8B					
T_2 94-2-04	1 (c)	6.782	10.658	9.3	28.9
	2 (p)	0.633	3.085		
T_2 94-2-05	3 (c)	$6.667 \cdot 10^6$	$12.917 \cdot 10^6$	6.7	23.0
	4 (p)	$0.448 \cdot 10^6$	$1.949 \cdot 10^6$		
T_2 94-2-11	5 (c)	$1.277 \cdot 10^6$	$4.495 \cdot 10^6$	6.1	9.4
	6 (p)	$0.078 \cdot 10^6$	$0.421 \cdot 10^6$		
Zheltozernoe 10 (original line)	7 (c)	$3.802 \cdot 10^6$	$12.034 \cdot 10^6$	48.7	37.2
	8 (p)	$1.853 \cdot 10^6$	$4.481 \cdot 10^6$		
Figure 9A					
T_3 94-2-11-2	1 (c)	$4.601 \cdot 10^6$	$7.055 \cdot 10^6$	13.9	11.6
	2 (p)	$0.638 \cdot 10^6$	$0.816 \cdot 10^6$		
T_2 94-2-11-3	3 (c)	$4.249 \cdot 10^6$	$6.829 \cdot 10^6$	7.9	10.4
	4 (p)	$0.336 \cdot 10^6$	$0.710 \cdot 10^6$		
Zheltozernoe 10 (original line)	5 (c)	$7.248 \cdot 10^6$	$21.939 \cdot 10^6$	47.8	33.4
	6 (p)	$3.464 \cdot 10^6$	$7.329 \cdot 10^6$		
Figure 9B					
T_3 94-3-08-2	1 (c)	$4.900 \cdot 10^6$	$8.845 \cdot 10^6$	6.8	13.9
	2 (p)	$0.331 \cdot 10^6$	$1.191 \cdot 10^6$		
T_3 94-3-08-3	3 (c)	$5.630 \cdot 10^6$	$10.256 \cdot 10^6$	4.3	8.7
	4 (p)	$0.243 \cdot 10^6$	$0.896 \cdot 10^6$		
T_3 94-3-08-1	5 (c)	$5.793 \cdot 10^6$	$8.656 \cdot 10^6$	13.5	12.6
	6 (p)	$0.782 \cdot 10^6$	$1.091 \cdot 10^6$		

Plant	Lane[1]	Estimated protein quantity[2]		Percent of undigested protein[3]	
		α-kafirin monomers	total	α-kafirin monomers	total
Zheltozernoe 10 (original line)	7 (c)	$10.090 \cdot 10^6$	$19.495 \cdot 10^6$	56.4	38.8
	8 (p)	$5.692 \cdot 10^6$	$7.570 \cdot 10^6$		

[1]c—control sample; p—pepsin treatment.

[2]Values are expressed as amount of dots (intensity × mm^2).

[3]Percentage from estimated protein quantity in undigested sample.

Table 2. Quantitative analysis of SDS-PAGE of total flour proteins from kernels of transgenic sorghum plants obtained by genetic transformation with *A. tumefaciens* GV3101/pNRKAFSIL.

Plant	Endosperm type	Protein digestibility (%)
Plants from T_1 and T_2 families		
T_1 94-2	Floury	85.4 c
T_1 94-6	Floury	85.2 c
T_2 94-2-05	Modified	74.1 b
T_2 94-2-11	Floury	90.7 cd
T_2 94-3-08	Vitreous, irregular	92.0 d
Zheltozernoe 10 (original non-transgenic line)	Vitreous	60.4 a
F		71.52**
Plants from T_3 families		
T_3 94-2-11-2	Modified	87.8 b
T_3 94-2-11-3	Modified	85.5 b
T_3 94-2-04-2	Modified	85.2 b
T_3 94-3-04-1	Floury	83.1 b
T_3 94-3-04-1	Modified	90.3 c
T_3 94-3-08-2	Vitreous, irregular	86.2 b
T_3 94-3-08-3	Vitreous, irregular	88.3 b
Zheltozernoe 10 (original non-transgenic line)	Vitreous	59.3 a
F		68.311**

Notes: Each value is a mean from two replications. Data followed by the same letter did not differ significantly ($P < 0.05$) from plant from the same group of families according to Duncan Multiple Range Test. Protein digestibility was calculated as percent ratio of difference between total estimated protein quantity in the control and digested sample to total estimated protein quantity in the control sample.

**Significant at $P < 0.01$.

Table 3. *In vitro* protein digestibility of sorghum flour from kernels of transgenic plants obtained by genetic transformation with *A. tumefaciens* GV3101/pNRKAFSIL.

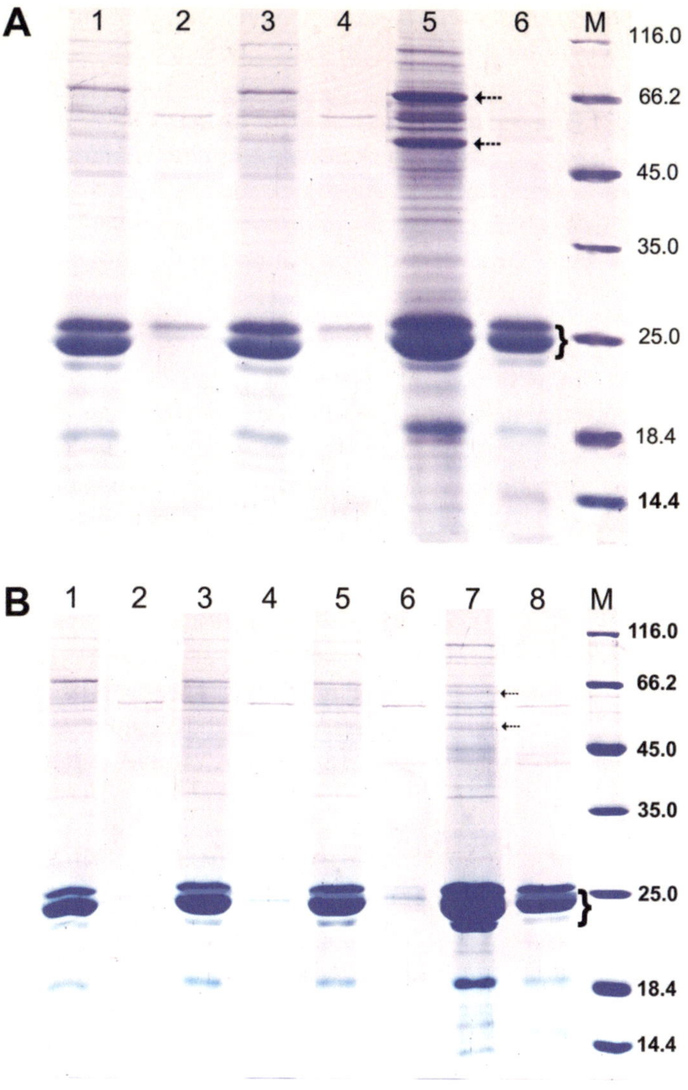

Figure 9. SDS-PAGE of endosperm proteins of kernels of transgenic sorghum plants from T_3 families #94-2-11 (with modified endosperm) and (with irregular vitreous endosperm) in reducing conditions. (A) 1, 2—#94-2-11-2; 3, 4—#94-2-11-3; 5, 6—original non-transgenic line Zh10; M—molecular weight markers (kDa). Dashed arrows indicate fraction of kafirin oligomers; brace—α-kafirin monomers. 1, 3, 5—control samples; 2, 4, 6—samples after pepsin digestion. (B) 1–6—Three individual plants from #94-3-08 family; 7, 8—original non-transgenic line Zh10. 1, 3, 5, 7—before and 2, 4, 6, 8—after pepsin digestion (published with the permission of the publishing house "Nauka").

Improved *in vitro* protein digestibility was observed also in plants from other T_3 families: #94-2-04, #94-3-04 and #94-3-08 (**Table 3**). In these plants, kernels had either floury or modified endosperm (#94-2-04-2; #94-3-04-1) or endosperm with irregularly developed vitreous layer (#94-3-08). Quantitative analysis showed that the level of digestibility of endosperm proteins in these plants was 83–90%, significantly differing from the digestibility of proteins in the original non-transgenic line.

Thus, the comparison of electrophoretic spectra of endosperm proteins before and after pepsin treatment showed a high level of kafirin digestibility in transgenic sorghum plants, harboring genetic construct for silencing of the γ-kafirin gene. Such electrophoretic spectra of digested endosperm proteins are not characteristic of ordinary sorghum cultivars obtained by classical breeding [40–42] except highly digestible sorghum mutant (*hdhl*) and its hybrids [37–39]. Apparently, a decrease in the level of γ-kafirin increases the digestibility of α-kafirins. This increase may be due to chemical reasons (reduction of polymerization) and/or physical reasons (change in the spatial arrangement of α-kafirins in the protein bodies that increase their availability to pepsin digestion).

Earlier it was reported on obtaining of transgenic sorghum plants carrying genetic constructs for silencing of γ- and α-kafirins, which were characterized by increased *in vitro* protein digestibility [25–27]. However, electrophoretic spectra of endosperm proteins after pepsin treatment were not shown in these studies. It should be noted also that in these studies improvement of kafirin digestibility was induced by complex genetic constructs that contained inverted repeats of several kafirin genes (δ2, γ1, γ2; or α1, δ2, γ1, γ2). These repeats were separated by the sequence of ADH1 intron, and the constructs were driven by the maize 19-kDa α-zein promoter [24–26]. In another work [27], the genetic construct included the complete sequence of the γ-kafirin gene, which was terminated by a nucleotide sequence of the self-cleaving ribozyme of tobacco ringspot virus that should destroy γ-kafirin mRNA. In our study [28], the effect was achieved by using a simpler genetic construct, containing inverted repeats of a short segment of the gene γ-kafirin (307 bp) separated by *ubi1*-intron gene, under the control of the constitutive *35S*-promoter, which allowed us to reach apparently rather high level of silencing of a target gene.

6. High-pressure liquid chromatography (HPLC) analysis of total amino acid content

An important feature of transgenic plants of sorghum and maize with silencing of prolamin genes is an increased proportion of essential amino acids in kernels, in particular the lysine proportion. Previously, this effect was observed in the silencing of genes of the main prolamin fractions: α-zein [22, 23] and α-kafirin [24, 27]. It was assumed that the suppression of the synthesis of these proteins, characterized by a low content of lysine, results in upregulating non-storage protein genes and appearance of lysine-rich proteins [23, 27].

In our experiments, the total amino acid content in the kernels of three transgenic plants from the T_2 generation with high *in vitro* protein digestibility: #94-2-11, #94-2-04 (both with modified endosperm) and #94-3-08 (with vitreous endosperm) was studied by using HPLC [28]. As can be seen from **Table 4**, content of a number of amino acids (leucine, proline, serine, isoleucine, histidine, tyrosine) and total amino acid content were significantly reduced in transgenic plants #94-2-04 (−40.2%, in comparison with the original non-transgenic line) and #94-3-08 (−22.8%). At the same time, the relative content of two major essential amino acids, lysine and threonine, significantly increased. Lysine proportion is increased by 1.6–1.7 times: from 1.54% of total amino acid content in the flour of original non-transgenic line Zh10 to 2.41–2.63% in transgenic plants #94-3-08 and #94-2-04, respectively, with vitreous and modified endosperms.

Amino acid	Control (Zh10)	T_2 94-2-11	T_2 94-2-04	T_2 94-3-08
Glu	2.84 ± 0.28	2.63 ± 0.15	**1.62 ± 0.07*** (−43.0%)	2.17 ± 0.08
Leu	1.72 ± 0.12	1.54 ± 0.08	**0.97 ± 0.04****(−43.6%)	**1.27 ± 0.03*** (−26.2%)
Ala	1.10 ± 0.11	1.02 ± 0.06	**0.65 ± 0.02*** (−40.9%)	0.85 ± 0,04
Pro	0.98 ± 0.07	0.85 ± 0.02	**0.58 ± 0.02**** (−40.8%)	**0.76 ± 0.02*** (−22.4%)
Asp	0.76 ± 0.03	0.74 ± 0.05	**0.44 ± 0.01**** (−42.1%)	0.57 ± 0.04
Phe	0.73 ± 0.01	0.59 ± 0.01	**0.39 ± 0.01*** (−46.6%)	0.49 ± 0.01
Ser	0.58 ± 0.03	0.52 ± 0.02	**0.35 ± 0.01**** (−39.7%)	**0.45 ± 0.01*** (−22.4%)
Val	0.57 ± 0.05	0.53 ± 0.02	**0.35 ± 0.01*** (−38.6%)	0.45 ± 0.01
Ile	0.48 ± 0.03	0.44 ± 0.02	**0.29 ± 0.01**** (−39.6%)	**0.36 ± 0.02*** (−25.0%)
Thr	0.40 ± 0.03	0.39 ± 0.01	**0.27 ± 0.00*** (−32.5%)	0.31 ± 0.08
Tyr	0.40 ± 0.02	0.38 ± 0.03	**0.24 ± 0.01**** (−40.0%)	**0.30 ± 0.02**** (−25.0%)
Arg	0.38 ± 0.02	0.33 ± 0.02	0.25 ± 0.02	0.32 ± 0,01
Gly	0.35 ± 0.03	0.34 ± 0.01	**0.25 ± 0.00*** (−28.6%)	0.31 ± 0.00
His	0.26 ± 0.01	0.23 ± 0.01	**0.15 ± 0.01**** (−42.3%)	**0.19 ± 0.01**** (−26.9%)
Lys	0.19 ± 0.04	0.23 ± 0.02	0.19 ± 0.00	0.22 ± 0.00
Cys/2	0.04 ± 0.00	0.04 ± 0.01	0.03 ± 0.01	0.04 ± 0.01
Met	0.03 ± 0.01	0.02 ± 0.00	0.02 ± 0.00	0.02 ± 0.00
Total	11.85 ± 0.86 c	10.86 ± 0.50 c	7.09 ± 0.24 a (−40.2%)	9.15 ± 0.23 b (−22.8%)
Lys (%)	1.54 a	2.14 b	2.63 c	2.41 bc
Thr (%)	3.37 a	3.59 a	3.86 b	3.42 a

Notes: Values are mean ± standard error from three replications. Data marked in bold differ significantly from the original non-transgenic (control) line Zh10 at $P < 0.05$ (*) and $P < 0.01$ (**) according to Student's T-test. Percentage of reduction is indicated in parenthesis.

Data for total amount of amino acid content and for percentage of lysine and threonine from the total amino acid content followed by the same letter did not differ significantly ($P < 0.05$) according to Duncan Multiple Range Test.

Table 4. Total amino acid content in kernels of transgenic sorghum plants obtained by genetic transformation with *A. tumefaciens* GV3101/pNRKAFSIL (g/100 g flour) (published with the permission of the publishing house "Nauka").

Such increase of the relative content of lysine and threonine in transgenic sorghum plants, coupled with a significant reduction of the total level of amino acids (**Table 4**), presumably was caused by decrease in the content of α-kafirins poor in lysine and threonine, whereas the synthesis of other proteins remained undisturbed. Accordingly, the relative proportions of lysine and threonine increased. The lower α-kafirins content in transgenic plants relative to the original non-transgenic line is clearly evident in the above electrophoresis photographs (**Figures 8** and **9**). Perhaps the suppression of the synthesis of γ-kafirin disrupts the formation of protein bodies and prevents the accumulation of α-kafirins, but does not affect the synthesis of other proteins richer in lysine and threonine.

7. Conclusion

Summarizing, using *Agrobacterium*-mediated genetic transformation with the strain carrying the genetic construct for silencing of γ-kafirin gene, we have obtained transgenic sorghum plants with significantly improved *in vitro* protein digestibility. The basis of such improved digestibility may be a reduction in the level of γ-kafirin, which causes formation of poorly digestible kafirin oligomers and development of vitreous endosperm. Further studies of these plants, including analysis of the expression of the genetic construct at the molecular level, will contribute to the understanding of regularities of endosperm development and possible use of these plants in sorghum breeding.

Obstacles along this path have both scientific reasons (instability of transgene expression, effects of transgenes on agronomically important traits) and social basis (public opposition to genetically modified plants). In future, to overcome public fears on "danger" of genetically modified organisms, sorghum plants with a modified synthesis of kafirins should be obtained by using marker-free technologies of genetic engineering or technologies of genome editing. With obtaining objective data from biosafety experiments, genetically modified sorghum plants with improved kafirin digestibility will be in demand on the market because they will combine favorable traits of sorghum (high grain productivity, resistance to drought stress) with a high nutritive value.

Acknowledgements

The work has been partially supported by Russian Foundation for Basic Researches, grants 10-04-00475, 13-04-01404, 16-04-01131.

Author details

Lev A. Elkonin[1]*, Julia V. Italyanskaya[1], Valery M. Panin[1] and Nikolai Yu. Selivanov[2]

*Address all correspondence to: lelkonin@gmail.com

1 Agricultural Research Institute of the South-East Region, Saratov, Russia

2 Institute of Biochemistry and Physiology of Plants and Microorganisms, Russian Academy of Sciences, Saratov, Russia

References

[1] Kamthan A, Chaudhuri A, Kamthan M, Datta A. Genetically modified (GM) crops: Milestones and new advances in crop improvement. Theoretical and Applied Genetics. 2016;**129**:1639-1655. DOI: 10.1007/s00122-016-2747-6

[2] Godwin ID, Williams SB, Pandit PS, Laidlaw HKC. Multifunctional grains for the future: Genetic engineering for enhanced and novel cereal quality. In Vitro Cellular & Developmental Biology—Plant. 2009;**45**:383-399. DOI: 10.1007/s11627-008-9175-5

[3] Katoch R, Thakur N. Advances in RNA interference technology and its impact on nutritional improvement, disease and insect control in plants. Applied Biochemistry and Biotechnology. 2013;**169**:1579-1605. DOI: 10.1007%2Fs12010-012-0046-5

[4] Younis A, Siddique MI, Kim C-K, Lim K-B. RNA interference (RNAi) induced gene silencing: A promising approach of Hi-tech plant breeding. International Journal of Biological Sciences. 2014;**10**:1150-1158. DOI: 10.7150/ijbs.10452

[5] El'konin LA, Domanina IV, Ital'yanskaya YV. Genetic engineering as a tool for modification of seed storage proteins and improvement of nutritional value of cereal grain. Agricultural Biology. 2016;**51**:17-30. DOI: 10.15389/agrobiology.2016.1.17eng

[6] Galili G, Amir R. Fortifying plants with the essential amino acids lysine and methionine to improve nutritional quality. Plant Biotechnology Journal. 2013;**11**(2):211-222. DOI: 10.1111/pbi.12025

[7] Blechl A, Lin J, Nguyen S, Chan R, Anderson OD, Dupont FM. Transgenic wheats with elevated levels of Dx5 and/or Dy10 high-molecular-weight glutenin subunits yield doughs with increased mixing strength and tolerance. Journal of Cereal Science. 2007;**45**:172-183. DOI: 10.1016/j.jcs.2006.07.009

[8] Leon E, Marin S, Gimenez MJ, Piston F, Rodriguez-Quijano M, Shewry PR, Barro F. Mixing properties and dough functionality of transgenic lines of a commercial wheat cultivar expressing the 1A61, 1D65 and 1Dy10 HMW glutenin subunit genes. Journal of Cereal Science. 2009;**49**:148-156. DOI: 10.1016/j.jcs.2008.08.002

[9] Gil-Humanes J, Piston F, Gimenez MJ, Martın A, Barro F. The introgression of RNAi silencing of γ-gliadins into commercial lines of bread wheat changes the mixing and technological properties of the dough. PLoS ONE. 2012;**7**(9):e45937. DOI: 10.1371/journal.pone.0045937

[10] de Mesa-Stonestreet NJ, Alavi S, Bean SR. Sorghum proteins: The concentration, isolation, modification, and food applications of kafirins. Journal of Food Science. 2010;**75**:90-104. DOI: 10.1111/j.1750-3841.2010.01623.x

[11] Henley EC, Taylor JRN, Obukosia SD. The importance of dietary protein in human health: Combating protein deficiency in Sub-Saharan Africa through transgenic biofortified sorghum. In: Taylor SL, editor. Advances in Food and Nutrition Research. Vol. 60. Burlington: Academic Press; 2010. pp. 21-52. DOI: 10.1016/S1043-4526(10)60002-2

[12] Oria MP, Hamaker BR, Shull JM. Resistance of sorghum α-, β- and γ-kafirins to pepsin digestion. Journal of Agricultural and Food Chemistry. 1995;**43**:2148-2153. DOI: 10.1021/jf00056a036

[13] Duodu KG, Taylor JRN, Belton PS, Hamaker BR. Factors affecting sorghum protein digestibility. Journal of Cereal Science. 2003;**38**:117-131. DOI: 10.1016/S0733-5210(03)00016-X

[14] Belton PS, Delgadillo I, Halford NG, Shewry PR. Kafirin structure and functionality. Journal of Cereal Science. 2006;**44**:272-286. DOI: 10.1016/j.jcs.2006.05.004

[15] Oria MP, Hamaker BR, Axtell JD, Huang CP. A highly digestible sorghum mutant cultivar exhibits a unique folded structure of endosperm protein bodies. Proceedings of the National Academy of Sciences of the United States of America. 2000;**97**:5065-5070. DOI: 10.1073/pnas.080076297

[16] Wu Y, Yuan L, Guo X, Holding DR, Messing J. Mutation in the seed storage protein kafirin creates a high-value food trait in sorghum. Nature Communications. 2013;**4**:2217. DOI: 10.1038/ncomms3217

[17] El Nour INA, Peruffo ADB, Curioni A. Characterization of sorghum kafirins in relation to their cross-linking behaviour. Journal of Cereal Science. 1998;**28**:197-207. DOI: 10.1006/jcrs.1998.0185

[18] Nunes A, Correia I, Barros A, Delgadillo I. Characterization of kafirin and zein oligomers by preparative dodecyl sulfate-polyacrylamide gel electrophoresis. Journal of Agricultural and Food Chemistry. 2005;**53**:639-643. DOI: 10.1021/jf049553+

[19] Duodu KG, Nunes A, Delgadillo I, Parker ML, Mills ENC, Belton PS, Taylor JRN. Effect of grain structure and cooking on sorghum and maize in vitro protein digestibility. Journal of Cereal Science. 2002;**35**:161-174. DOI: 10.1006/jcrs.2001.0411

[20] Kumar T, Howe A, Sato S, Dweikat I, Clemente T. Sorghum transformation: Overview and utility. In: Paterson AH, editor. Genetics and Genomics of the Saccharinae (Plant Genetics and Genomics: Crops and Models). Vol. 11. New York, NY: Springer; 2012. pp. 205-221. DOI: 10.1007/978-1-4419-5947-8_10

[21] Wu E, Lenderts B, Glassman K, Berezowska-Kaniewska M, Christensen H, Asmus T, Zhen S, Chu U, Cho M-J, Zhao Z-Y. Optimized *Agrobacterium*-mediated sorghum transformation protocol and molecular data of transgenic sorghum plants. In Vitro Cellular & Developmental Biology—Plant. 2014;**50**:9-18. DOI: 10.1007/s11627-013-9583-z

[22] Segal G, Song R, Messing J. A new opaque variant of maize by a single dominant RNA-interference-inducing transgene. Genetics. 2003;**165**:387-397. Available from: http://www.genetics.org/content/165/1/387

[23] Huang S, Frizzi A, Florida CA, Kruger DE, Luethy MH. High lysine and high tryptophan transgenic maize resulting from the reduction of both 19- and 22-kD α-zeins. Plant Molecular Biology. 2006;**61**:525-535. DOI: 10.1007/s11103-006-0027-6

[24] da Silva LS, Jung R, Zhao ZY, Glassman K, Grootboom AW, Mehlo L, O'Kennedy MM, Taylor J, Taylor JRN. Effect of suppressing the synthesis of different kafirin sub-classes on grain endosperm texture, protein body structure and protein nutritional quality in improved sorghum lines. Journal of Cereal Science. 2011;**54**:160-167. DOI: 10.1016/j.jcs.2011.04.009

[25] da Silva LS, Taylor J, Taylor JRN. Transgenic sorghum with altered kafirin synthesis: Kafirin solubility, polymerization and protein digestion. Journal of Agricultural and Food Chemistry. 2011;**59**:9265-9270. DOI: 10.1021/jf201878p

[26] Grootboom AW, Mkhonza NL, Mbambo Z, O'Kennedy MM, da Silva LS, Taylor J, Taylor JRN, Chikwamba R, Mehlo L. Co-suppression of synthesis of major α-kafirin sub-class together with γ-kafirin-1 and γ-kafirin-2 required for substantially improved protein digestibility in transgenic sorghum. Plant Cell Reports. 2014;**33**:521-537. DOI: 10.1007/s00299-013-1556-5

[27] Kumar T, Dweikat I, Sato S, Ge Z, Nersesian N, Elthon T, Bean S, Ioerger BP, Tiley M, Clemente T. Modulation of kernel storage proteins in grain sorghum (*Sorghum bicolor* (L.) Moench). Plant Biotechnology Journal. 2012;**10**:533-544. DOI: 10.1111/j.1467-7652.2012.00685.x

[28] Elkonin LA, Italianskaya JV, Domanina IV, Selivanov NY, Rakitin AL, Ravin NV. Transgenic sorghum with improved digestibility of storage proteins obtained by agrobacterium-mediated transformation. Russian Journal of Plant Physiology. 2016;**63**:678-689. DOI: 10.1134/S1021443716050046

[29] de Barros EG, Takasaki K, Kirleis AW, Larkins BA. Nucleotide sequence of a cDNA clone encoding γ-kafirin protein from *Sorghum bicolor*. Plant Physiology. 1991;**97**:1606-1607. DOI: 10.1104/pp.97.4.1606

[30] Gelvin SB. *Agrobacterium* virulence gene induction. In: Wang K, editor. *Agrobacterium* Protocols. Vol. 1. Methods in Molecular Biology. Vol. 343. Totowa, NJ: Humana Press Inc.; 2006. pp. 77-84

[31] Chilton M-D, Currier TC, Farrand SK, Bendich AJ, Gordon MP, Nester EW. *Agrobacterium tumefaciens* DNA and PS8 bacteriophage DNA not detected in crown gall tumors. Proceedings of the National Academy of Sciences of the United States of America. 1974;**71**:3672. Available from: http://www.pnas.org/content/71/9/3672

[32] Zhao ZY, Cai TS, Tagliani L, Miller M, Wang N, Pang H, Rudert M, Schroeder S, Hondred D, Seltzer J, Pierce D. *Agrobacterium*-mediated sorghum transformation. Plant Molecular Biology. 2000;**44**:789-798. DOI: 10.1023/A:1026507517182

[33] Elkonin LA, Pakhomova NV. Influence of nitrogen and phosphorus on induction embryogenic callus of sorghum. Plant Cell Tissue and Organ Culture. 2000;**61**:115-123. DOI: 10.1023/A:1006472418218

[34] Emani C, Sunilkumar G, Rathore KS. Transgene silencing and reactivation in sorghum. Plant Science. 2002;**162**:181-192. DOI: 10.1016/S0168-9452(01)00559-3

[35] Laemmli UK. Cleavage of structural proteins during the assembly of the head of bacteriophage T4. Nature. 1970;227:680-685. DOI: 10.1038/227680a0

[36] Wu Y, Holding DR, Messing J. γ-Zeins are essential for endosperm modification in quality protein maize. Proceedings of the National Academy of Sciences of the United States of America. 2010;**107**:12810-12815. DOI: 10.1073/pnas.1004721107

[37] Weaver CA, Hamaker BR, Axtell JD. Discovery of grain sorghum germplasm with high uncooked and cooked *in vitro* protein digestibility. Cereal Chemistry. 1998;**75**:665-670. DOI: 10.1094/CCHEM.1998.75.5.665

[38] Tesso T, Ejeta G, Chandrashekar A, Huang CP, Tandjung A, Lewamy M, Axtell J, Hamaker BR. A novel modified endosperm texture in a mutant high-protein digestibility/high-lysine grain sorghum (*Sorghum bicolor* (L.) Moench). Cereal Chemistry. 2006;**83**:194-201. DOI: 10.1094/CC-83-0194

[39] Aboubacar A, Axtell JD, Huang CP, Hamaker BR. A rapid protein digestibility assay for identifying highly digestible sorghum lines. Cereal Chemistry. 2001;**78**:160-165. DOI: 10.1094/CCHEM.2001.78.2.160

[40] Nunes A, Correia I, Barros A, Delgadillo I. Sequential in vitro pepsin digestion of uncooked and cooked sorghum and maize samples. Journal of Agricultural and Food Chemistry. 2004;**52**:2052-2058. DOI: 10.1021/jf0348830

[41] Elkonin LA, Italianskaya JV, Fadeeva IY, Bychkova VV, Kozhemyakin VV. *In vitro* protein digestibility in grain sorghum: Effect of genotype and interaction with starch digestibility. Euphytica. 2013;**193**:327-337. DOI: 10.1007/s10681-013-0920-4

[42] Wong JH, Marx DB, Wilson JD, Buchanan BB, Lemaux PG, Pedersen JF. Principal component analysis and biochemical characterization of protein and starch reveal primary targets for improving sorghum grain. Plant Science. 2010;**179**:598-611. DOI: 10.1016/j.plantsci.2010.08.020

Chapter 7

Vascular Tissue Development and Regeneration in the Model Plant *Arabidopsis*

Ewa Mazur and Jiří Friml

Additional information is available at the end of the chapter

http://dx.doi.org/10.5772/intechopen.69712

Abstract

Development of vascular tissue is a remarkable example of intercellular communication and coordinated development involving hormonal signaling and tissue polarity. Thus far, studies on vascular patterning and regeneration have been conducted mainly in trees—woody plants—with a well-developed layer of vascular cambium and secondary tissues. Trees are difficult to use as genetic models, i.e., due to long generation time, unstable environmental conditions, and lack of available mutants and transgenic lines. Therefore, the use of the main genetic model plant *Arabidopsis thaliana* (L.) Heynh., with a wealth of available marker and transgenic lines, provides a unique opportunity to address molecular mechanism of vascular tissue formation and regeneration. With specific treatments, the tiny weed *Arabidopsis* can serve as a model to understand the growth of mighty trees and interconnect a tree physiology with molecular genetics and cell biology of *Arabidopsis*.

Keywords: *Arabidopsis*, vascular tissue, vascular cambium, secondary xylem, auxin, auxin transporters, cellular polarity, PIN proteins

1. Introduction

Various species and systems were used for the analysis of vascular tissue [1–5]; however, *Arabidopsis* has been demonstrated to be the most suited plant for studies of molecular biology and developmental genetics due to its model status [6]. This in combination with rapid and effective induction system for vascularization [7, 8] with features, such as functioning vascular cambium and secondary vascular tissues found in woody plants, established in *Arabidopsis* will provide decisive advantage over previous approaches.

In this review, we summarize information concerning secondary vascular tissue development in *Arabidopsis* including cambium ontogenesis and xylogenesis, with accompanying changes in auxin distribution, directionality of its flow, and cellular polarity defined by auxin transporters (PIN family proteins), which have been indicated to be involved in regulation of vascular tissue patterning and regeneration [7–9].

2. Secondary vascular tissues in woody plants

Vascular tissue is a well function conducting system typical for all woody plants, among them in trees. In young plants, characteristic primary tissues such as procambium, primary xylem, and primary phloem, develop. During the secondary growth, vascular tissue undergoes the transition from primary into the secondary vascular patterning. Vascular cambium, secondary xylem, and secondary phloem form a closed ring on the stem circumference. They are arranged in the radial rows as a consequence of periclinal divisions of cambial cells [10–12]. The secondary growth is mostly characteristic for all woody plants, and production of the secondary vascular tissues is an important developmental feature of the plants [12, 13].

2.1. Vascular cambium

Vascular cambium plays a crucial role in the secondary growth and vascular tissue patterning in woody plants [14, 15]. Activity and functioning of vascular cambium decide about the amount of the secondary phloem and xylem, which are produced outward and inward the vascular cambium, respectively [12, 13]. This meristematic tissue is built from two types of cells: ray cambial cells producing secondary rays—transverse conducting system in tissues—and fusiform cambial cells, producing elements of the longitudinal conducting systems in woody plants. Characteristic feature of the vascular cambium is intrusive growth of the fusiform cambial cells and their periclinal divisions [13, 16, 17]. The intrusive growth is restricted to the ends of growing cells, when two neighboring cells grow in opposite directions. Periclinal divisions of cambial cells decide about production of cambial derivatives and secondary tissue element differentiation.

Vascular cambium is the tissue very much sensitive to mechanical injuries, such as wounding or grafting. However, it can easily regenerate under suitable conditions. It has been experimentally shown that cambium regeneration is mostly dependent on the tensile stress and pressure. Results obtained by Brown [18] indicate that cambium activity, cell divisions, and xylem formation can be easily affected by the pressure externally implied to the cambial strips. It is also documented in *in vivo* experiments with the wounding stems of *Larix europea* that regeneration of this meristematic tissue can dynamically progress under the tensile stress and pressure implication. In such cases, even low pressure (25 kPa) implied to cambium in wounded areas decides about its very rapid regeneration. Lack of the mechanical factors leads to abundant callus tissue production [19, 20].

It has been postulated that appropriate functioning of vascular cambium and its cyclic activity, i.e., periclinal divisions during the seasons, is strictly correlated with auxin signaling and auxin responses [21–23]. From the studies on *Populus tremula* L. × *Populus tremuloides*, it appears that the highest auxin concentration is found in the layer of cambium. Auxin plays here a key role in the regulation of cambial cell divisions and elongation [23]. According to the results obtained with the vascular cambium in *Pinus sylvestris*, thickness of the cambial layer is directly dependent on the auxin concentration in the tissue, which stimulates frequency of divisions [21, 22]. The highest activity of vascular cambium is found at the beginning of the vegetative seasons, in early spring, which is correlated with the first periclinal divisions of the cambial cells, production of new cambial derivatives, and their differentiation into new secondary vascular tissues. Otherwise, the lowest cambial activity in winter, during the dormant period, is strictly correlated with decreasing both temperature and hormonal levels in cambium layer [24]. Periclinal divisions are limited and almost completely stop; thus, the vasculature is not produced during this time. However, it was experimentally shown that such situation could be easily reversed after exogenous auxin application [25]. As a consequence, activity of cambium and periclinal divisions of cambial cells can be resumed by auxin. Thus, from all the studies on the woody plants, it appears that elevated auxin response in cambial cells as well as fluctuations of auxin (maxima/minima) in cambium plays decisive role in seasonal nature of the trees, for example, switching on/off dormant periods [21, 22]. Changes of the cell wall components [26–28] and gene expression during the cyclic activity of vascular cambium [29, 30], correlated with the rapid cytoskeleton rearrangement in differentiating cells [31], indicate that this meristematic tissue plays a crucial role in the secondary growth of woody plants and decides about their adaptation to variable environmental conditions.

2.2. Secondary xylem

In the most typical form, secondary xylem, also called a wood, is found in stems and roots of the woody plants. The secondary xylem, a longitudinal conducting system in trees, develops from the cambial derivatives, which during the maturation process is differentiated into elements of the wood-like vessels, fibers, and tracheids [13, 15].

Vessels of the secondary xylem form strands parallel to the longitudinal axis of the organs—stems or roots. Every vessel strand is consisted of single vessel elements, the so-called vessel members, connected with each other by open perforation plates localized on their apical-basal ends [12]. It is postulated that direction of vessel differentiation is dependent on direction of auxin flow. Thus, in nondisturbed stems, vessels developed according to the polar auxin transport (PAT) in the apical-basal direction whereas in incised organs—according to newly established direction of auxin flow—circumventing the wounded regions. Correlations between auxin flow and vasculature patterning were experimentally documented in woody plants after wounding [32] as well as nonwoody models [4, 5, 8]. Characteristic feature for all types of vessels (primary protoxylem, metaxylem, and secondary xylem vessels) is the secondary cell wall. Different patterning of the secondary cell wall is realized during vessel maturation process and depends on the type of vessel [14]. During the maturation process, protoplasts of differentiating vessels disappeared. Frequently, the lumen of the vessel members is enlarged in comparison to other tracheary elements, mainly in a wood of such species as *Fraxinus excelsior*, *Quercus borealis*, or *Ulmus americana* [13]. In many cases, length of the vessel members is different than the length of the fusiform cambial cells from which they developed. Interestingly, longitudinal vessel strands change the orientation to the longitudinal axis of stems. Such fluctuations are observed as the wavy grain patterning of wood in many trees [33].

Fibers of the secondary xylem are recognized as one of the longest tracheary elements, characterized by the tapered cell ends and reduced lumen. As a consequence of their intensive intrusive growth, fibers could be even few times longer than the fusiform cambial cells and their derivatives. Particular type of the woody fibers is the so-called gelatin fibers, developed as a layer of the reaction wood in many deciduous as well as coniferous trees, i.e., *Populus* sp. or *Picea* sp. Inner layer of secondary cell wall of these fibers is built mainly from cellulose. The presence of the callose and callose-like cell wall components plays here an important role in mechanical properties of the wood [34].

Tracheids, other tracheary elements of secondary xylem, are nonperforated, long cells with the bordered pits. Dependently on the type of a wood, tracheids are classified as (1) vessel-like tracheids arranged in longitudinal, similar to vessels conducting strands, commonly found in *Carpinus* sp., *Ulmus* sp., *Acer* sp., or *Tilia* sp.; (2) tracheids differentiated around the vessels with enlarge lumen, adjacent to them, and strictly surrounding; and (3) fiber-like tracheids [35]. The last of them develop as a conducting and storage water system but also play mechanical functions and in some species could be the main component of the softwood of gymnosperms [36]; they are found also in some angiosperms, i.e., in *Populus* sp.

Besides of the dead, water-conducting elements of the secondary xylem mentioned above, in many cases secondary vascular tissue of woody plants is compound with the xylem parenchyma cells, which remain alive for a long time to finally die in the programmed cell death (PCD) process [12].

Thorough knowledge about the genetic and molecular mechanisms involved in vascular tissue functioning, development, and regeneration is eagerly expected. Different molecular components involved in the determination of developmental plasticity of cambial cells have been searched for with special interest focused on the key regulators of vascularization. Genes involved in auxin response, auxin signaling pathways, and tissue and cellular polarity during vascular tissue development induced in vascular cambium should be extensively studied for detailed characterization of this process.

2.3. *Arabidopsis* as a nonwoody plant example for vascular tissue formation

Since many years, *Arabidopsis* is nominated as a good model for studies of vascular tissue formation, because under suitable conditions, *Arabidopsis* can undergo secondary growth in hypocotyls, when enlarged layer of secondary xylem develops during xylogenesis [37]. Xylogenesis in hypocotyls is comparable to xylogenesis in roots. Development of secondary xylem is divided here into two phases: the early phase, xylem is building from vessels and numerous parenchyma cells, and in the second, later phase, also called xylem expansion, enlarged amount of xylem elements develops mainly vessels and fibers [37, 38]. It is well documented that vascular tissue develops not only in hypocotyls [37] but also in the matured inflorescence stems [39–41], in their basal parts [42–45]. With the use of *Arabidopsis*, the correlation between auxin signaling and tissue polarity has been intensively studied and modeled [46–49]. However, because of lack of some important vasculature features such as a variety of typical phenotype features and functional cambium, these models could not be used for full analysis and description of vascular tissue development and compared to the analogical process in trees. For example, in both *Arabidopsis* models mentioned above, rays were not found. Also intrusive growth typical for fusiform cambial cells was not confirmed. Finally, impressive variety of tracheary elements, among them tracheids, commonly found in trees, were not observed in *Arabidopsis* mature stems and hypocotyls, which underwent secondary growth [50].

In contrast, in the *Arabidopsis* inflorescence stems stimulated mechanically by an artificial weight, the transition from primary to secondary tissue architecture leads to the development of all vasculature features mimicking secondary vascular tissues in woody plants. According to the new approach, immature inflorescence stems (9–10 cm tall) were firstly decapitated with the sharp razor blade (shoot apex and flowers were removed). Next, the artificial weight (2.5 g) was applied to the decapitated apical parts of the stems. Stems were additionally supported by a wood stick to avoid their bending. Importantly, the axillary buds grown above the leave rosettes were not removed, thus remaining the natural source of endogenous auxin. This experimental approach has been extensively described [7, 8]. It was speculated that the weight carried by the stem serves as a mechanically stimulated signal for wood formation [7, 8, 42, 51]. According to Ko and coauthors [42], mechanical stimulation of immature inflorescence stems of *Arabidopsis* increases polar auxin transport and promotes the secondary growth. It allows designing *Arabidopsis* as a full "tree-like" system. Moreover, the secondary vascular tissues develop in a very short time, namely, in 6 days [7], which is much faster than in hypocotyls [37, 38] or mature inflorescence stems of *Arabidopsis* [44, 45].

In the created *Arabidopsis* "tree-mimicking" model, the development of variety of vascular cambium phenotypes is the most spectacular. According to the obtained results, both types of cambial cells develop: (1) ray cambial cells, very short and almost round cells arranged in single-row rays mimicking transverse conducting system in woody plants, and (2) fusiform cambial cells, long, tapered-end cells, characterized by the intrusive growth and periclinal divisions, play here an important role in secondary tissue element differentiation. The phenomenon of intrusive growth of fusiform cambial cells is described for the first time in the mechanically stimulated *Arabidopsis* stems (**Figure 1A–D**), not found in the previously analyzed models. Neighboring cells start their growth in the opposite directions, but the growth is restricted to the tips of the cells, which slide along the radial cell walls and provide the elongation of the fusiform cambial cells (**Figure 1A, C**, and **D**).

New approach, based on mechanical stimulation of the immature inflorescence stems of *Arabidopsis* [7, 8, 42, 51], is expected to elucidate the phenomenon of vascular tissue formation and regeneration at cellular and molecular level—processes commonly studied in woody plants, but not fully explained yet, because of some experimental and environmental difficulties in these plants [52]. Thus, *Arabidopsis* comes out as a good model system for vascular tissue patterning.

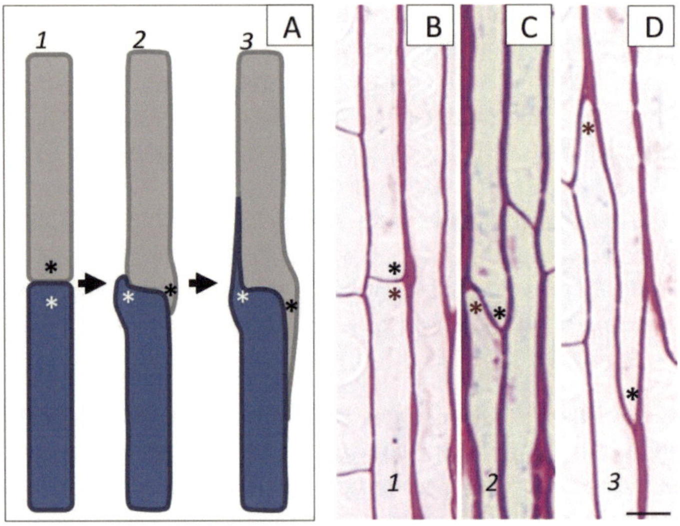

Figure 1. Intrusive growth of the fusiform cambial cells in mechanically stimulated *Arabidopsis* stems. (A) Schematic visualization of two intrusively growing cells ((1) two neighboring cells with still non-intrusively growing ends, (2) beginning of the intrusive growth of the opposite ends, (3) advanced intrusive growth along the radial cell walls of the cells). Asterisks indicate neighbor cells and their intrusively growing ends. (B–D) Intrusively growing fusiform cambial cells in temporal steps corresponding to situation visualized in (A). (B–D = longitudinal tangential sections through *Arabidopsis* stems; Poly/Bed 812 resin sections stained with the Periodic Acid—Schiff's (PAS reaction); bar, 10 μm).

3. Vascular tissue development and regeneration in mechanically stimulated inflorescence stems of *Arabidopsis*

In this paragraph, we will describe in detail the transition from primary to secondary tissue architecture in inflorescence stems of *Arabidopsis* as the important step in obtaining a suitable model for secondary vascular tissue analysis, following the temporal and spatial changes during vascular cambium ontogenesis, xylem formation, and vascular tissue regeneration in weight-induced *Arabidopsis*.

3.1. Ontogenesis of vascular cambium

Ontogenesis of vascular cambium is correlated with temporal and spatial changes on the stem circumference. Usually, formation of a closed ring of cambium is preceded by dedifferentiation of parenchyma cells into cambial cells and the so-called interfascicular cambium development. This process is commonly observed in young woody plants during their secondary growth [12]. It has been confirmed by histological analyses that the first dedifferentiated parenchyma cells are localized next to the vascular bundles in the early stages of the interfascicular cambium development [12]. With the time, the regions of dedifferentiating parenchyma cells are extended and finally enclosed as continuous ring on the stem circumference. The mechanism of these changes is still not clarified. The basic question is which of the cellular events trigger the parenchyma cell dedifferentiation?

http://dx.doi.org/10.5772/intechopen.69712

In mechanically stimulated *Arabidopsis* stems, vascular cambium develops from fascicular cambium and interfascicular cambium bands (**Figure 2**). Fascicular cambium develops as a primary meristematic tissue in vascular bundles, localized in the inner parts of immature stems characterized by primary tissue architecture. The vascular bundles are separated by interfascicular parenchyma bands with nonpericlinally dividing parenchyma cells (**Figure 2A** and **B**). Outside these regions, few layers of cortex and single layer of the epidermis are situated. Middle parts of the stems consisted of the enlarged, thin-cell wall pith parenchyma cells. One- or few-layer supporting tissue with characteristic thick-cell wall interfascicular fibers plays mechanical function in immature stems (**Figure 2B**). In immature stems of *Arabidopsis*, 6 days after weight application, the architecture of the basal parts of such stems diametrically changes. At the beginning of the secondary growth, interfascicular cambium develops in the interfascicular regions of stems, as a consequence of parenchyma cell dedifferentiation (**Figure 2C** and **D**). Interestingly, the most inner layer of interfascicular parenchyma cells dedifferentiates into the interfascicular cambium. Typically, it is a single layer of parenchyma

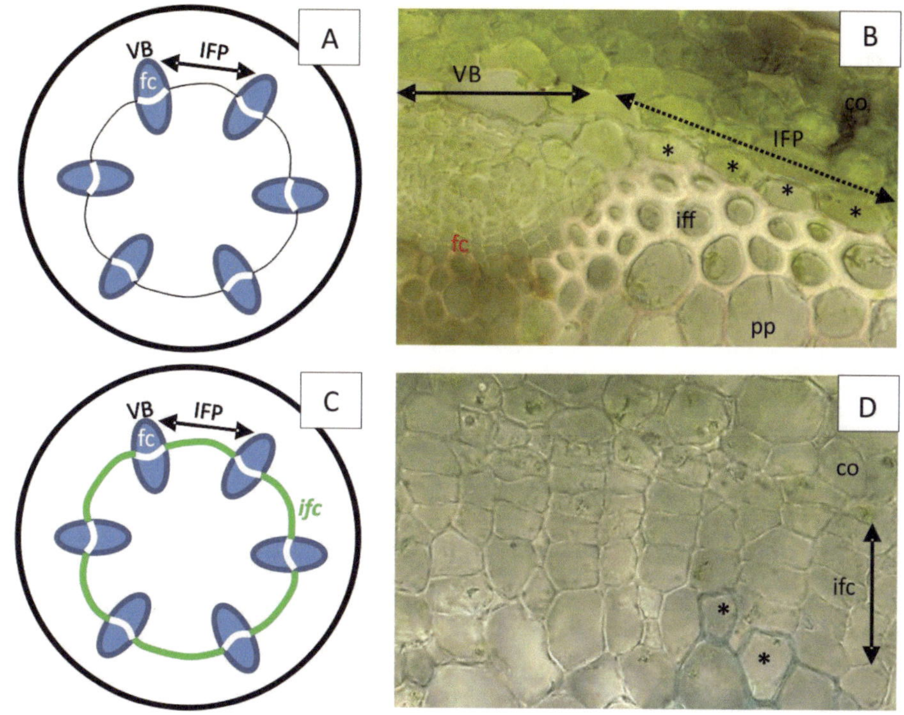

Figure 2. Transition from primary to secondary tissue architecture in weight applied inflorescence stems of *Arabidopsis*. (A) Schematic visualization of the tissue arrangement in immature inflorescence stems of *Arabidopsis*. (B) Cross section through the basal parts of stem with the primary tissues—vascular bundles with fascicular cambium are separated by the interfascicular parenchyma. The most inner layer of interfascicular parenchyma cells will dedifferentiate into cambial cells (asterisks). (C) Schematic visualization of the secondary vascular tissues with closed ring of vascular cambium on stems circumference. (D) Layer of periclinally dividing interfascicular cambial cells as a part of the ring of vascular cambium. The first developed vessels are indicated by asterisks (lignin in the secondary cell walls stained with 0.05% Toluidine Blue O aquatic solution). B and D, handily made cross sections through basal part of stems (VB, vascular bundle; IFP, interfascicular parenchyma; *iff*, interfascicular fibers; *co*, cortex; *pp*, pith parenchyma; *fc*, fascicular cambium; *ifc*, interfascicular cambium); bars, 20 μm.

cells localized between vascular bundles. Finally, during the transition from the primary to the secondary tissue architecture of *Arabidopsis* stems, fascicular and interfascicular cambium forms fully enclosed ring of vascular cambium on stem circumference (**Figure 2D**).

The whole process of cambium ontogenesis is strictly correlated with such cellular events as elevated auxin response in interfascicular parenchyma, polarity of parenchyma cells dedifferentiating into the cambium, their periclinal divisions, and changes of their cell wall components [7]. The most spectacular seems to be correlations between auxin response and tissue polarity during cambium ontogenesis in analyzed *Arabidopsis* stems. Already in the first few days, auxin concentration distinctly arises in the dedifferentiating parenchyma cells. At the early stages of the interfascicular cambium development, maximum auxin concentration is detected in parenchyma cells localized in the nearest neighborhood of vascular bundles, whereas in the later stages of this process, the zone of the cells with elevated auxin response is gradually extended toward the middle parts of the interfascicular regions, in the next few days after weight application [7]. Polarity of the interfascicular parenchyma was monitored by the PIN-FORMED1 (PIN1) protein localization in differentiating cells. The PINs are well-known auxin transport proteins involved in the cellular efflux of auxin and polar auxin transport in plant tissues [53]. In many developmental processes, the establishment of local PIN-dependent auxin gradient in cells is strictly correlated with cellular divisions and developmental reprogramming [54, 55].

During analyzed process of cambium ontogenesis, tissue polarity is rapidly established in *Arabidopsis* stems. Amazingly, polarity of interfascicular parenchyma is indicated by polar localization of PIN1 auxin transport protein, which localizes at the basal plasma membranes of differentiating cells [56]. It has been documented that the protein appears in the basal plasma membranes of dedifferentiating parenchyma cells, not previously found in parenchymatic cells of immature mechanically noninduced *Arabidopsis* stems [7]. Moreover, both of the events—elevated auxin response and tissue polarization—are accompanied by periclinal divisions of the parenchyma

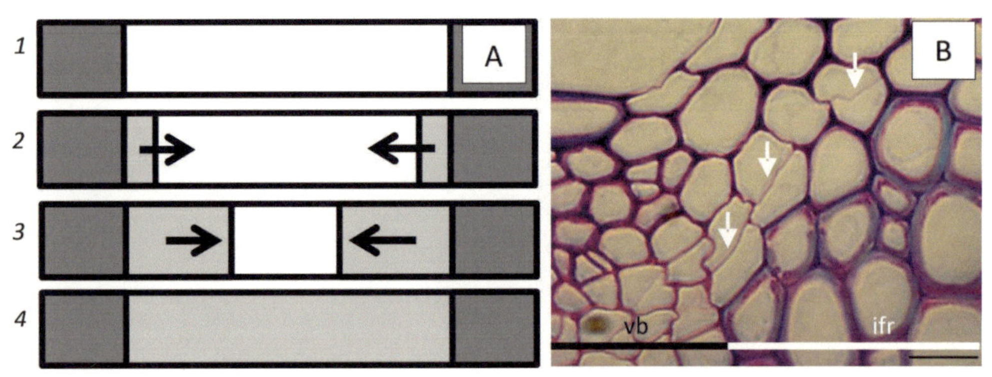

Figure 3. Periclinal divisions of interfascicular parenchyma cells and interfascicular cambium development in weight-induced *Arabidopsis* stems. (A) Schematic visualization of the temporal changes in interfascicular parenchyma regions with gradually extended zone of periclinally dividing cells (1–4 = four steps of the changes from vascular bundles, dark grey; zone of dividing parenchyma cells, grey; nondivided parenchyma cells, white; arrows indicate the direction of changes). (B) Periclinal divisions of the interfascicular parenchyma cells in the neighborhood of the vascular bundle (arrows); Poly/Bed 812 resin section stained with the periodic acid-Schiff's (PAS reaction); *vb*, vascular bundle; *ifr*, interfascicular region; bar, 20 μm.

cells (**Figure 3**). Divisions are temporarily correlated with the cellular events mentioned above and maintained in space. Namely, the first periclinally divided cells appear in the neighborhood of vascular bundles (**Figure 3A** and **B**), but later the zone of dividing cells slowly extends toward the middle part of interfascicular regions. In consequence, parenchyma cells dedifferentiate into cambial cells, which definitely changes architecture of the interfascicular regions and decides about development of the interfascicular cambium (**Figure 3A**). According to the obtained results, it is tempting to conclude that auxin plays the most important role during cambium ontogenesis in *Arabidopsis* stems. Auxin seems to be a primary signal for cellular fate reprogramming and a crucial clue for stimulation of the dedifferentiational process in the interfascicular parenchyma zones.

In the described model, vascular cambium could be classified as "functioning" meristematic tissue, which actively produces cambial derivatives. Differentiation of cambial derivatives is a consequence of numerous periclinal divisions of fusiform cambial cells. Finally, the maturation of the cambial derivatives into secondary vascular tissue elements supported functionality of this meristematic tissue in the present model. The sequence of the changes could be useful for all comparative analysis of the cambium ontogenesis and xylogenesis both in *Arabidopsis* model system and the analogical mechanisms studied in woody plants. Thus, the mechanically stimulated *Arabidopsis* model with fully functional cambial meristem could help us in addressing the elusive vascularization mechanisms observed in the woody plants.

3.2. Secondary xylem formation in *Arabidopsis* stems

Reprogramming of the gene expression that accompanies xylogenesis and transdifferentiation of mesophyll cells into tracheary elements was extensively studied in *in vitro* cultures of zinnia (*Zinnia elegans*) [57, 58]. However, the lack of the cambium stage in this experimental system prevents us from deciphering the role of cambium in wood formation. Temporal gene expression pattern accompanies dedifferentiation of cambial cells into cambial derivatives, but their maturation into different types of tracheary elements is poorly characterized. Thus, numerous efforts have been focused on the identification of master regulatory genes required for this transition and revealing the key components of the vascular-differentiation-involved genetic network [48, 59].

In *Arabidopsis* "cambial" model, vascular cambium reveals basic features of functioning cambium important in the following stages of xylogenesis. Periclinal divisions of the fusiform cambial cells lead to the development of secondary xylem derivatives in the early stage of xylogenesis. Changes in later stages of xylogenesis are correlated with maturation of cambial derivatives into tracheary elements and secondary vascular xylem development (**Figure 4**). During this process, such recognizable tracheary elements as vessels, fibers, or tracheids develop and create the layer of secondary xylem. Vessels are easily recognized, because of some diagnostic features such as secondary cell wall and open perforation plates on the opposite ends of the vessel members (**Figure 4C**). Vessels are arranged in threads of longitudinal strands in the vascular tissue. Amazingly, in the present *Arabidopsis* model, impressive variety of tracheary elements is detected, not previously documented in analyzed hypocotyls [37, 38] or adult stems of *Arabidopsis* [39–41].

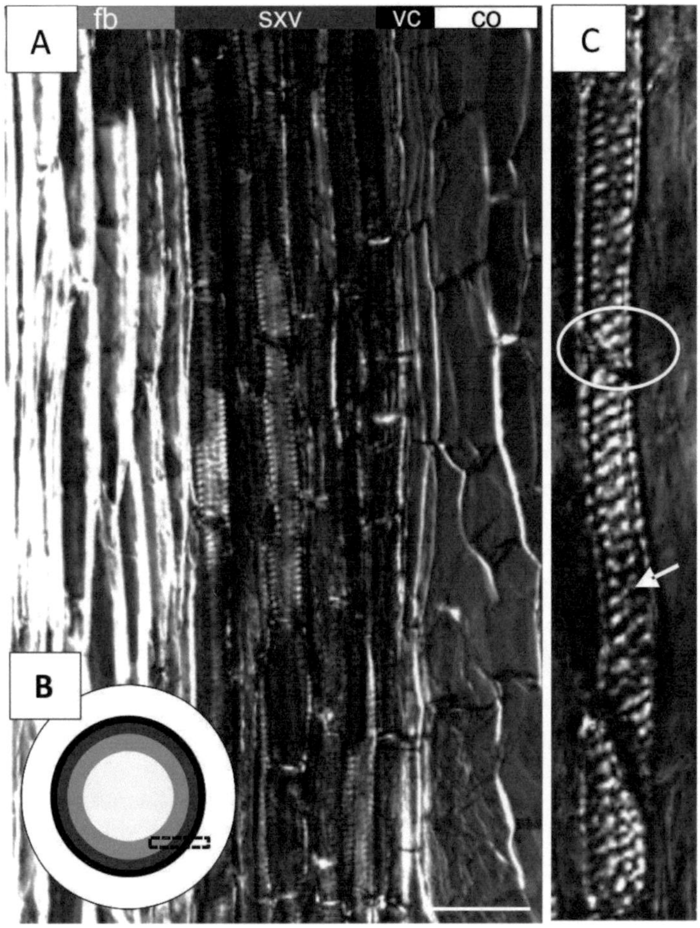

Figure 4. Secondary xylem in the weight stimulated stems of *Arabidopsis*. (A) Secondary xylem elements, like vessels and fibers, are produced from cambial derivatives after numerous periclinal divisions of fusiform cambial cells. Cortex parenchyma is visible outside the secondary vascular tissues. (B) Schematic visualization of the tissue arrangement in stem and localization of the tissues showed in (A) is indicated by the square. (C) Vessel strand developed parallel to longitudinal axis of stem. Characteristic patterning of the secondary cell wall (arrow) and perforation plates developed on the opposite apical-basal ends of neighboring vessel members (circle) determines the most diagnostic features for this type of tracheary elements (A and C, bright-field images in a confocal laser-scanning microscope; *fb*, fibers; *sxv*, secondary xylem vessels; *vc*, vascular cambium; *co*, cortex); bars, 50 μm (A); 20 μm (C).

Patterning of vascular tissue and variety of tracheary elements developed as a dynamically operating water-conducting system and was extensively studied in the woody plants [13, 14]. However, mechanism regulating xylogenesis at cellular and molecular levels remains unclear, and many questions are unanswered. For example, differentiation of tracheids as a type of tracheary elements commonly found in trees, but for the first time detected in mechanically stimulated *Arabidopsis*, led to important conclusions about the involvement of the artificial weight in wood formation. Following stages of xylogenesis involving formation of the variety of tracheary elements, such as recognized tracheids, will be helpful in future analysis.

3.3. Regeneration of vascular tissue in wounded *Arabidopsis* stems

In 1981, Sachs postulated canalization hypothesis according to which vasculature patterning is based on the positive feedback loop between auxin flow and cellular polarity. Consequently, in the primary uniform tissue, cellular auxin transporters emerge as the so-called auxin channels that transport the hormone through the tissue in the polar direction. Emergence of auxin channels is correlated with establishment of cellular polarity inside these specific auxin transport routs. Finally, new vessels develop directly along the auxin channels. Canalization hypothesis is strongly supported by many classical experiments with the incised plants, i.e., by wounding or grafting, which shows that emergence of auxin channels is correlated with increased auxin response and tissue repolarization [1, 2, 4, 5]. It is well documented that initially broadly elevated auxin response in wounded tissues is gradually restricted to narrow auxin channels, in which auxin level is still very high [4]. The obtained results showed that patterning of vascular tissue, explicitly visible during regeneration and new vasculature development, is dependent on new ways of canalized auxin flow.

Well-functioning vascular cambium plays the most important role for the secondary growth in the woody plants, both secondary xylem formation and stem thickness [14, 21, 22, 60]. Many results revealed an important role for this meristematic tissue during vasculature regeneration process. For decades analysis of vascular patterning and incised vascular cambium regeneration was restricted mainly to trees [61–63] because these woody plants undergo secondary growth with enlarged amount of secondary xylem (wood) and active cylinder of vascular cambium [64]. Studies were based mainly on the histological analysis, thus limited only to the final effects of regeneration. Thus, it was impossible to analyze vasculature regeneration, including vascular cambium, on the cellular and molecular levels. Some experimental studies on trees showed that in the wounded areas, the cambium and vascular tissue regenerate very fast both *in vivo* [19, 20] and *in vitro* [25, 65, 66]. Regeneration is accompanied by numerous anticlinal divisions of cambial cells and their dynamic intrusive growth [19, 20, 64], which finally leads to the reconstruction of vasculature and new vessel patterning in the incised regions [25, 65]. In some instances, when the auxin flow is locally reversed, the so-called circular vessels develop [32, 67, 68]. In the nondisturbed woody plants, circular vessels are often found in branch junctions, above the axillary buds [68], whereas in incised plants, after transversal cuts and exogenous auxin application to stem segments, in wounded regions [32, 67]. Accordingly, circular vessels occur in the form of rings and are presumably induced as a consequence of the circular auxin flow and the establishment of the circular polarity of individual cells that dedifferentiated into this type of vessels [67]. Thus, according to Sachs and Cohen [67], circular vessels develop as a response of individual cells to the auxin flux rather than to the high local auxin concentration. In nonwoody dicotyledonous plants characterized by primary tissue architecture, such as *Phaseolus vulgaris*, *Pisum sativum*, or *Coleus* sp., vasculature is regenerated directly from dedifferentiated parenchyma cells [1–5]. New vessels are arranged either around the wound according to the presumable new auxin flow [69] or form the so-called bypass strands directly through the wound [3] or bridges between the neighboring vascular bundles [70]. Lack of the vascular cambium in the studied nonwoody plants restricted a detailed analysis of regeneration of this meristematic tissue and cellular events accompanying this process. Therefore in the used models, the most intriguing questions are still remained of answer: (1) what is the role of vascular cambium in vascular tissue regeneration?, (2) which of the cellular events are temporary correlated with the vascular cambium regeneration?, (3) is vascular cambium regeneration mediated by the canalization process? Full verification of the postulated canalization hypothesis and identification of the molecular mechanisms accompanied vascular tissue regeneration are still limited.

Because of the difficulties in using woody plants as a convenient model system [52], mechanisms of cambium regeneration are still poorly understood. With the *Arabidopsis* "cambium" model, it is now possible to monitor vascular tissue regeneration with all cellular events accompanying this process. Thus, in control conditions, i.e., in nonincised stems, polar auxin flow is in the direction from apical to basal part of stems, and according to this flow, new vasculature develops. Otherwise, in incised stems (i.e., wounded stems), polar auxin transport is disturbed; thus, new ways of auxin flow are established. As a consequence, new vessel strand arrangement is changed, because the new vasculature likely developed according to new directions of auxin cell-to-cell transport (**Figure 5**). In wounded *Arabidopsis* stems, threads of new vessel strands develop above or around a wound (**Figure 5A** and **B**, respectively). Interestingly, vessels above a wound regenerated faster, in the first days after wounding (DAW) (2 and 3 days), whereas vessel around a wound differentiated in the next few days, beginning the day 4 and circumventing the incised areas. They developed from cells after their numerous, uneven divisions,

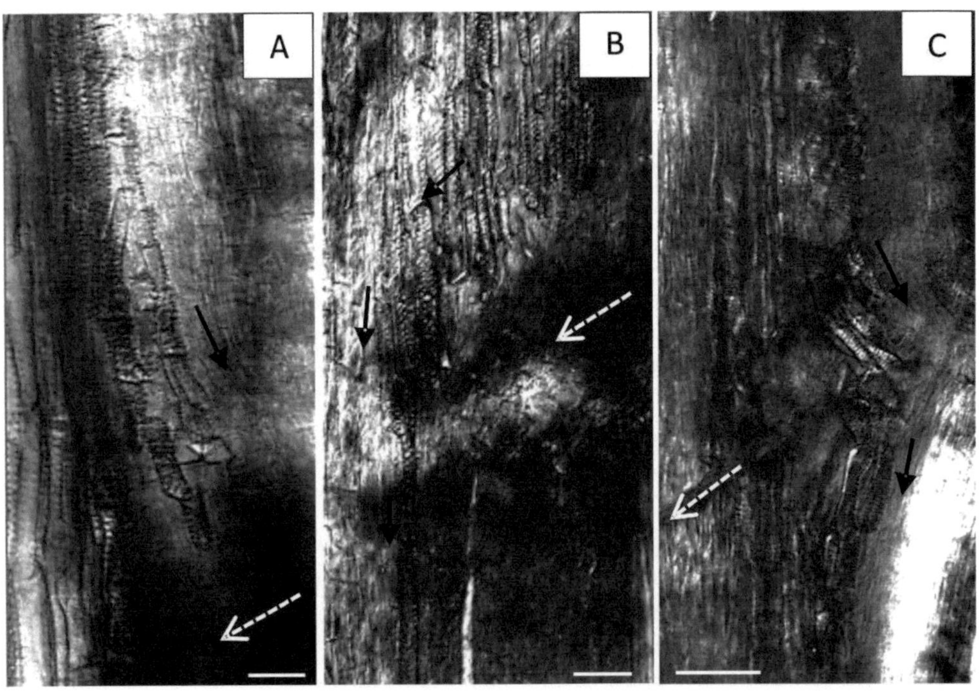

Figure 5. Paths of vessel regeneration in wounded *Arabidopsis* stems. (A) Threads of short vessel members developed above a wound. (B) Vessel strands regenerated around a wound. (C) Vessel "bypass" strands reconstructed partially from the callus tissue developed inside the wound. Arrows indicate regenerated vessel strands. Broken arrows mark places of the wound; bright-field images in a confocal laser-scanning microscope; bars, 50 μm.

what is commonly observed in the wounded tissue. Differentiating vessels were visualized by the activity of the *AtHB8* gene, which belongs to HD-ZIP III family [71, 72]. The *AtHB8* is positively regulated by auxin, and its extensive activity in wounded regions during vascular tissue regeneration suggested that *AtHB8* might play a crucial role in the vasculature development [71, 72]. The last observed way of vasculature regeneration is correlated with callus differentiation (**Figure 5C**). Namely, in wounded areas vessels develop from previously proliferated callus tissue cells. Such vessels often create the type of "bypass" strands extending above and below the transversal incision.

Regeneration of vascular tissue in wounded *Arabidopsis* stems is accompanied by temporal and spatial changes following new vessel development. New vessel strands regenerated in the incised regions around a wound develop as a consequence of cambial cell regeneration. Longitudinal continuum of vascular cambium is disturbed after the transversal cut. In such experimental system, rapid auxin response is found as a primary signal of the regeneration. Merely at the first day after incision, elevated auxin concentration is observed above a wound and in the next few days also around a wound [8]. Vasculature regeneration is strictly correlated with tissue repolarization and establishment of new polarity in neighborhood of the wound. Tissue repolarization always preceded emergence of PIN1-positive auxin channels (**Figure 6**). As a consequence, layer of new vessels develops around a wound, and the regenerated vasculature becomes enlarged

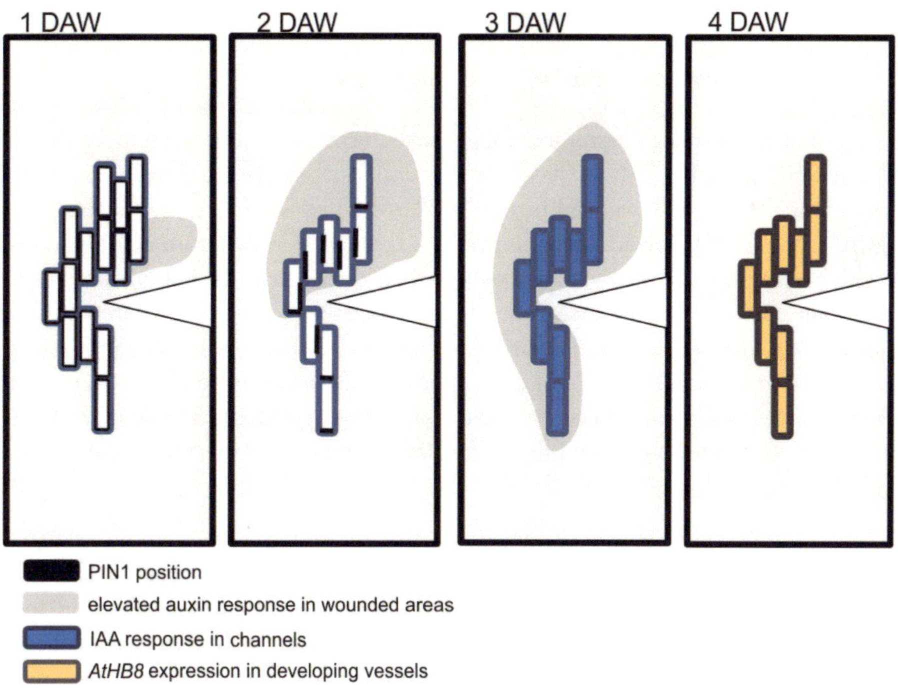

Figure 6. Schematic visualization of the temporal and spatial changes during vascular tissue regeneration in following days after wounding (DAW): re-localization of PIN1 in cells around a wound (1 DAW); establishment of new position of PIN1 at cellular plasma membranes (2 DAW); increased auxin concentration in auxin channels (3 DAW); expression of *AtHB8* gene in differentiating vessels (4 DAW).

in the days following the incision. Analysis of regeneration process in incised *Arabidopsis* stems strongly supported canalization hypothesis. Emergence of new vasculature is correlated here with elevated auxin response and changed polarity in auxin channels, from which new vessel strands develop in the wounded areas.

4. Role of plant hormone auxin and auxin transporters in vascular tissue development

Auxin is regarded as a multifunction plant hormone, which plays a fundamental role in developmental processes during organo- and morphogenesis. Auxin is a primary signal in regulation of many cellular processes, which control oriented divisions, cell elongation, or differentiation. At last, auxin is a key hormonal factor inducing vascularization—vascular tissue development, patterning, and regeneration. Polar auxin transport (PAT) manifested as physiological, basipetal direction of auxin flow represents a unique mechanism specific to plants. The cellular and molecular action of this process, explained in the chemiosmotic model, is based on auxin influx and efflux carriers, namely, AUX and PIN proteins, which actively participate in the cell-to-cell hormone transport [73–75]. The local auxin accumulation, its minima and maxima, or the so-called gradients in tissues are precisely controlled by this process.

4.1. Auxin as a primary signal inducing vascularization

The role of auxin as a primary signaling cue in vascularization has been widely discussed for decades. Experiments with radioactively labeled auxin show its maximum concentration in the meristematic tissues such as cambium [22, 57] and in adjacent cambial derivatives, differentiating into xylem [76]. Periodic fluctuation of auxin concentration in cambium influences the frequency of cambial cell divisions, production of cambial derivatives, and secondary vascular tissues. Disturbance of these correlations leads to many defects in cambium functioning and xylem formation. Using transgenic lines of *Arabidopsis*, elevated auxin response is easily found just in the cambial cells of both types of cambia (**Figure 7**). Auxin concentration is very high in the fascicular cambium bands, primary meristematic tissue in the vascular bundles (**Figure 7**), as well as in the interfascicular vascular cambium on the stem circumference (**Figure 7**).

From the experimental studies on the vascularization *in vitro*, it appears that parenchyma callus tissue is the most convenient for the analysis. Previously uniform callus can form vascular tissue bands or groups of vessels differentiation. However, the process can be realized only in the sufficiently thick callus tissue. It is shown that differentiated xylem in surrounded by cambium-like cells, which additionally are able to produce phloem elements in the inner callus regions. Auxin-dependent vascularization is also shown in the studies with young *Syringa* sp. stems [77]. Combination of auxin and sucrose decides about the induction of vascularization in the axillary buds *in vitro*. Moreover, dependent on the hormone and sucrose concentration, varied vascular tissues develop.

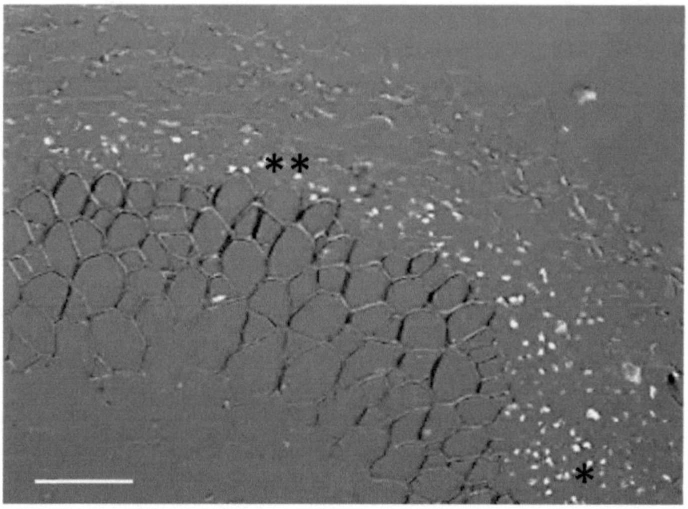

Figure 7. Elevated auxin concentration in cambium of non-incised *Arabidopsis* control stems. Increased auxin level in fascicular cambium (vascular bundle, asterisk) and in the interfascicular cambial cells (double asterisks); cross section through the basal part of Col-0 stem; confocal laser-scanning microscope; IAA immunolocalization-staining with the polyclonal anti-IAA antibody (white color in cambial cells; dilution 1:500); bar, 50 μm.

Several reports discussed auxin as a specific morphogenetic signal triggering cell fates during vascular tissue development and its maturation [78]. Locally created centers characterized by elevated auxin response become more competent for auxin flow through primarily uniform tissues. Auxin waves created in plant organs as a specific system of hormonal information that decide about realization of many developmental programs in plants, among them cambial activity and differential cambial responses [79, 80]. Thus analogically, gradual emergence of auxin channels and gradually narrowing auxin flow finally results in vascular strand differentiation. In other words, canalized auxin flux determined the paths of new vasculature development.

The canalization-predicted vasculature formation is especially observed during regeneration process, in new regenerated vessels after incision [1, 2, 4, 5, 8, 81]. Particularly important contributions to the role of auxin in the vascular tissue differentiation brought studies on *Pisum* sp. [1, 2]. According to all experiments performed by Sachs, vascularization depends on the polar auxin transport, and new vascular band induction depends on the auxin concentration and polarity. Moreover, the early stages of vascular band differentiation are related to the canalization of the polar auxin flow. A key role of auxin in promotion of canalized flow by itself and transport channels formation is commonly accented. However, the feedback mechanism between auxin flow, polarity, and vessel formation as a response to concentration gradients or directional auxin fluxes remains unclear [82, 83].

4.2. Role of auxin transporters in cellular and tissue polarity

The positive feedback loop between polar auxin flow and the polar, subcellular localization of the PIN-FORMED (PIN) auxin transport proteins [56] that, in turn, determine the auxin flow directionality is widely studied [53, 54, 84–86]. Many developmental processes, such as early embryogenesis or plant organ initiation, are strictly correlated with the establishment of local PIN-dependent auxin gradients that precede cell divisions and differentiation [54, 55, 87]. The expression of auxin efflux carrier genes, like *PIN1*, *PIN2*, *PIN3*, *PIN4*, and *PIN7* was found to peak at the inflorescence stems of *Arabidopsis* during their maturation and secondary vascular tissue development. Changes in PIN localization and tissue polarity in response to auxin that are presumably related to the directional vascular tissue patterning have been observed and modeled [4, 5, 46, 88]. Moreover, in wounded pea or bean epicotyls, the PIN polarity was gradually rearranged marking the position of differentiating vessel strands [4, 5]. Emergence of auxin channels is here visualized by *PIN1* expression of the cellular auxin transporters. In *Arabidopsis* model with mechanically stimulated inflorescence stems, the subcellular PIN1 position was gradually stabilized and restricted only to cell sides in a first few days after weight application, along the presumable direction of the auxin flow [8]. The auxin-dependent canalization is strongly supported by studies on leaf vein patterning and on the role of the genes encoding the auxin response factor *MONOPTEROS* (*MP*) and *PIN1* [47]. Dynamic expression of both of the genes and gradual establishment of polarized PIN1 protein localization indicates the direction auxin flow during the vascular tissue patterning in analyzed leaves [47]. Moreover, the other *Arabidopsis* gene *GNOM/EMB30*, which affects apical-basal position of PIN1, seems to be required for regulation of the coordinated tissue polarity [6].

The role of auxin transporters in vascular tissue patterning is clearly visible in wounded inflorescence stems of *Arabidopsis*, during vascular cambium regeneration [8]. Rapid tissue repolarization indicated by reposition of PIN1 at cellular plasma membranes of differentiating cells is emphasized. Dynamic temporal changes in tissue polarity are correlated with varied auxin response and its accumulation above and around a wound. Whereas auxin concentration arises in few hours after wounding, maximum of auxin levels is established at auxin channels and preceded establishment of new polarity in wounded areas of *Arabidopsis* stems. Cellular auxin transporters are characterized with changed position of PIN1 proteins. Thus, direction of auxin flow through the auxin channels is precisely determined. Both of the events are strictly correlated with each other and play a decisive role in vascular tissue development.

4.3. Auxin signaling pathways in vascular tissue patterning

Two related protein families—Aux/IAA and ARFs—are well-known key regulators of auxin-modulated gene expression and act in the TIR1-mediated signaling pathway [89, 90]. Members of ARF family share the characteristic arrangement of a highly conserved DNA-binding domain near the N-terminus, which appear to be capable to auxin response elements (AuxREs)—short conserved sequences (TGTCTC) that have been shown to be essential for auxin regulation of auxin-inducible genes [6]. It is likely that the ARF proteins are strongly involved in the vascularization downstream proteolytic SCF^{TIR1} complex machinery [80]. In support of this, an increased level of ARF transcripts was differentially regulated during the secondary growth, and three of them (ARF2, ARF4, and ARF5) had the most dramatic expression changes, indicating their putative roles in apical-basal signaling and xylogenesis [6, 42]. On the other hand, the *AUXIN-RESISTANT 1* (*AXR1*) gene is required for normal *TIR1* function and, when mutated, changes the stabilization dynamics of the Aux/IAA proteins [71]. Mutations in the TIR1/AFBs make Aux/IAA proteins insensitive to auxin and can therefore keep ARF transcription factors and auxin signaling repressed. Thus, ARFs together with Aux/IAA proteins constitute a central mechanism in auxin signaling during plant development [91].

Auxin besides regulating a gene expression by the TIR1/AFB pathway can also inhibit internalization of the PIN proteins by a feedback regulation [92]. The underlying perception and signaling mechanism is unclear, but it does not involve transcription regulation and is distinct from the TIR pathway [93]. It may relay on the Auxin Binding Protein 1 (ABP1) since the ABP1 overexpressors increase the PIN internalization and mutations in the auxin binding pocket of ABP1 make the ABP1 effect on PIN internalization auxin-insensitive [88]; however, due to unreliable loss-of-function data [94, 95], this issue requires further clarification.

The identification of spatiotemporal gene expression pattern and the key components of auxin signaling pathway/pathways will greatly contribute to understanding of the molecular mechanisms involved in auxin-induced regeneration switch in cambial cells. In addition, the knowledge on genetic factors, such as ARFs, AFBs involved in the SCF^{TIR1} auxin receptor complex, PIN auxin efflux transporters, or AtHB family of early vascularization markers determining developmental plasticity of cambial cells, can be useful in genetic improvement of woody plants for environment and biotechnology purposes.

5. Genetic control of vascularization processes

Numerous genes differentially regulated during vascularization in woody plants are expected to be identified with the use of obtained *Arabidopsis* model. Many of the genes were indicated to be involved in auxin responses implicating auxin engagement in regulation of vascular tissue development and patterning [42, 43], and the same is expected for vascular tissue regeneration. Experimental data have proven a key role of auxin in variety of developmental processes [80]; however, the molecular, auxin-mediated mechanism involved in vasculature regeneration remains mostly unknown.

As indicated recently, transcription factor genes promoting secondary growth induction [42, 43] can be applied in genetic transformation to improve our knowledge on xylogenesis and regeneration capacity of woody plants. Several regulatory genes, like *NAC*-domain genes or key regulators of SAM, shoot apical meristem, organization (*WUS, STM, WOX, CLVs*), are closely associated with the vascular tissue formation. It was reported that two *NAC*-genes, *NAM* and *CUC*, and *ANT*—a cell proliferation marker—play potential roles in vascular tissue differentiation and function [96]. *ANT* regulates organ growth through the maintenance of meristematic tissue activity.

The expression of the homeobox genes (*AtHBs*) is highly increased during xylem production and regarded as a positive regulator of the activity of procambial and cambial cells to differentiate [71, 72]. Baima et al. [72] reported ectopic expression of *AtHB-8* gene during vascular regeneration after wounding in *Arabidopsis*. Extensive activity of this gene was found in the regenerated tissues, suggesting intensive transcriptional reprogramming during new vessel development. In the model with functioning vascular cambium, expression of *AtHB-8* is observed in differentiating cambial derivatives, in early stages of their maturation into the vessels (**Figure 8**).

Homeostasis of vascular cambium with its non-disturbed functionality plays an important role in the vascularization [59]. However, genetic control of vascular cambium activity is poorly characterized. The role of the leucine-rich repeat receptor-like kinase (LRR-RLK) families in regulation of this lateral meristem homeostasis, underlying the role of *CLV1* (*CLAVATA1*)—well-known apical meristem marker and PXY (phloem intercalated with xylem)—a cambium-specific receptor-like kinase, in this process is suspected [59]. Two receptor-like kinases MOL1 (more lateral growth1) and RUL1 (reduced in lateral growth1) identified as opposing regulators of cambium activity were also reported [59]. Recently, the role of the homeobox transcription factor WOX4 (wuschel-related homeobox 4), as an essential cambium regulator positively regulated by *PXY*, has been revealed [59, 97]. Since many years the correlations between the cytoskeleton dynamic (both actin filaments and microtubules) and activity and functioning of vascular cambium [31, 37, 98] have been widely discussed. It was postulated that changes in

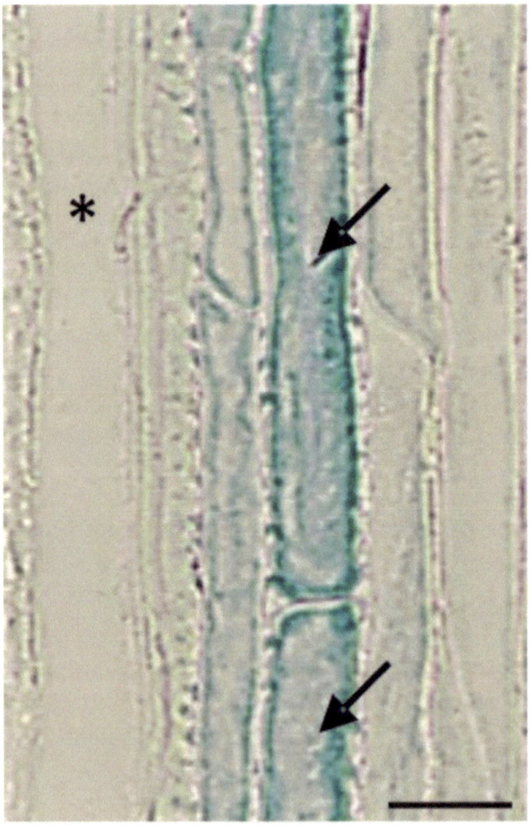

Figure 8. The *AtHB8* gene expression in differentiating vessels. Longitudinal tangential section through the basal part of stem. Expression of *AtHB8* in differentiating vessels (arrows); asterisk indicates maturated vessel (*AtHB8:GUS* transgenic line; LR white resin section; bar: 20 μm).

microtubule orientation might be specific cellular marker for cambial cells in the stage of their differentiation into tracheary elements. Analyses of the following steps of this process revealed the dynamic changes in the microtubule orientation in the differentiating cambial cells [98].

The most important role in the whole vascular-formation and regeneration machinery is played by auxin response genes referred to as early/primary auxin response genes. Three major classes of these genes, *Aux/IAA*, *SAURs*, and *GH3*, are characterized by the so-called auxin response elements (AuxREs), the TGTCTC-containing auxin response promoter elements. The genes are specifically induced by auxin, which may rapidly regulate their transcription. Accordingly, when auxin concentration in cells is low, auxin response genes are repressed. Oppositely, when auxin concentration is elevated, transcription of the genes is activated in a few minutes [99], and ARF transcription factors are binding to its AuxRE target sides. It is discussed that dependently on the tissue and developmental stages of the plants, tissue-specific expression of both ARF and Aux/IAA proteins could regulate the transcription of different sets of genes for different developmental processes.

Studies in the *Populus tomentosa* [29] have shown that different groups of genes are activated dependent on the stage of the secondary vascular tissue development. Thus, in the cambial zone, high expression of the *ARK1* gene was observed. It was suggested that *ARK1* regulates cambium activity including cambial cell divisions and differentiation of the cambial derivatives [30]. On the contrary, during the secondary xylem formation, expression of the group of genes responsible for the tracheary element differentiation and maturation, such as *SND1* or *NST1* and *NST3*, was reported [100, 101].

It is widely postulated that important role in the vascularization is played also by gene expression that accompanies this process, like *PLT* (*PLETHORA*) or *TCH* (*TOUCH*) genes encoding calcium-binding proteins. Whereas *PLT*s regulate de novo shoot regeneration in *Arabidopsis* by controlling successive developmental events, *TCH* genes (*TCH2* and *TCH4*) strongly induced by mechanical stimuli like touch and wind [42] may be involved in the signal transduction and secondary xylem formation. Otherwise, two of the *VASCULAR-RELATED NAC-DOMAIN* genes (*VND6* and *VND7*) are reported as positive xylem vessel differentiation regulators in both *Arabidopsis* and poplar [48]. Thus, the elusive mechanism for auxin-regulated vascular tissue patterning might be a part of the extensive genetic network including several hormonal signaling pathways and dynamic spatiotemporal switched on/off gene expression. Deciphering of these yet unknown relationships will help to translate the mechanisms regulated in vascular tissue development and regeneration in woody plants.

6. Emergence of *Arabidopsis* as a good model system to study the vascular tissue formation and regeneration processes

For a long time, the herbaceous plant *Arabidopsis* was postulated as a major model system for developmental plant biology due to its some important features, such as a short generation time and relatively small size of fully sequenced genome [52]. It was used as a modern research tool for different molecular studies, especially thanks to large mutant and transgenic line collections. Finally, in many studies *Arabidopsis* was also postulated as a perfect plant for secondary growth analysis [37–39, 44, 45, 50, 102, 103].

Although the vascularization and regeneration in *Arabidopsis* have been previously analyzed [3, 81], the role of vascular cambium in these processes has never been addressed. Only just modified and established method to obtain stems with closed vascular cambium rings and some typical features mimicking vascular tissue in trees [7, 51], such as secondary rays and intrusively growing cambial cells or tracheids—unusual example of tracheary elements found in woody plants—gives infinite possibilities to analyze secondary vascular tissue development and provides decisive advantage over previous approaches with the use of *Arabidopsis* stems. This strategy is particularly suited to elucidate different molecular mechanisms and other molecular components involved in auxin-mediated responses, canalization of auxin flow, and cellular polarity, underlying determination of plant developmental plasticity.

The availability of numerous genetic and molecular tools in *Arabidopsis* will provide clarifying picture of the process of vasculature formation as well as reconstruction after wounding and decisively extend the knowledge on molecular control of spatiotemporal vasculature patterning and regeneration both in vivo and in vitro. Resolving the involvement, the mechanism of canalization process will contribute to explain molecular mechanisms involved in vasculature regeneration and provide a useful model for further studies.

7. Conclusions

Specifically manipulated *Arabidopsis* represents a good system for the analysis of vascularization machinery that typically occurs in trees. Obtained results revealed that this process is accompanied with cellular events following cambium ontogenesis, xylem formation, and regeneration: (1) elevated auxin concentration in tissues, (2) new polarity establishment, (3) reposition of PIN1 proteins at plasma membranes of differentiating cells, (4) cellular divisions, and finally (5) cambium and vascular tissue development or regeneration.

Knowledge about the molecular mechanisms regulating vascular tissue development in trees is incomplete, and such studies can take full practical advantages from a recently proposed new approach. Temporal analysis and experiments in trees are hampered mainly because of variability in environmental conditions, their long life cycle, and restricted amount of transgenic lines and mutants available.

Currently, new insights into the vascular tissue formation problematics can be obtained by using *Arabidopsis thaliana* (L.) Heynh. This modest plant commonly used as a genetic model can be also modified to become a perfect experimental "tree-like" model, with a closed ring of functional cambium and secondary vascular tissues with complexity of phenotype features fully comparable to woody plants. This new approach promises to identify additional genetic and molecular components involved in vascular tissue functioning, including the insights into the role of cambial cells in this process in woody plants. Extensive studies with the use of

Arabidopsis model system allow obtaining a more complete data about vasculature development and regeneration under more controlled experimental conditions.

Interdisciplinary scientific approach with the use of *Arabidopsis* as the "tree-like" model promises to pioneer new, original, and valuable information about the mechanisms regulating vascularization and allows defining molecular factors of the auxin-dependent machinery involved in this process. This creates a unique opportunity to compare the processes in *Arabidopsis* with analogical situation found in trees. There is a hope that the obtained knowledge could be applied practically. Plant life conditions are closely related to dynamic environmental and climate changes, which is a global problem of the last decades. Among the most endangered ecosystems are forests, which are shrinking very fast not only because of environmental changes but mostly because of the non-sustainable economic exploitation. Thus obtaining additional options to study typical tree features including their regenerative abilities is important to design efficient strategies for sustainable wood production.

Acknowledgements

This publication was funded by the European Research Council under the European Union's Seventh Framework Programme (FP7/2007-2013)/ERC grant agreement n° 282300.

Author details

Ewa Mazur[1]* and Jiří Friml[2]

*Address all correspondence to: ewa.mazur@us.edu.pl

1 Department of Cell Biology, Faculty of Biology and Environmental Protection, University of Silesia in Katowice, Poland

2 Institute of Science and Technology, Klosterneuburg, Vienna, Austria

References

[1] Sachs T. The induction of transport channels by auxin. Planta. 1975;**127**:201-206

[2] Sachs T. The control of the patterned differentiation of vascular tissues. Advances in Botanical Research. 1981;**9**:151-262

[3] Flaishman MA, Loginovsky K, Lev-Yadun S. Regenerative xylem in inflorescence stems of *Arabidopsis thaliana*. Journal of Plant Growth Regulation. 2003;**22**:253-258

[4] Sauer M, Balla J, Luschnig C, Wiśniewska J, Reinöhl V, Friml J, Benková E. Canalization of auxin flow by Aux/IAA-ARF-dependent feedback regulation of PIN polarity. Genes and Development. 2006;**20**:2902-2911

[5] Balla J, Kalousek P, Reinöhl V, Friml J, Procházka S. Competitive canalization of PIN-dependent auxin flow from axillary buds controls pea bud outgrowth. Plant Journal. 2011;**65**:571-577

[6] Berleth T, Mattsson J, Hardtke CS. Vascular continuity, cell axialisation and auxin. Plant Growth Regulation. 2000;**32**:173-185

[7] Mazur E, Kurczyńska EU, Friml J. Cellular events during interfascicular cambium ontogenesis in inflorescence stems of *Arabidopsis*. Protoplasma. 2014;**251**:1125-1139

[8] Mazur E, Benkova E, Friml J. Vascular cambium regeneration and vessel formation in wounded inflorescence stems of Arabidopsis. Scientific Reports. 2016;**69**:33754. DOI: 10.1038/srep33754

[9] Scarpella E, Marcos D, Friml J, Berleth T. Control of leaf vascular patterning by polar auxin transport. Genes and Development. 2006;**20**:1015-1027

[10] Newman IV. Pattern in the meristems of vascular plants. I. Cell partition in living apices and in the cambial zone in relation to concept of initial cells and apical cells. Phytomorphology. 1956;**6**:1-19

[11] Savidge RA. The role of plant hormones in higher plant cellular differentiation. II. Experiments with the vascular cambium, and sclereids and tracheid differentiation in the pine, *Pinus contorta*. Histochemical Journal. 1983;**15**:447-466

[12] Evert RF. Esau's Plant Anatomy: Meristems, Cells, and Tissues of the Structure, Function, and Development, 3rd ed. New Jersey: Wiley; 2006

[13] Larson PR. The Vascular Cambium. Development and Structure. Berlin: Springer-Verlag; 1994

[14] Philipson WR, Ward JM, Butterfield BG The Vascular Cambium: Its Development and Activity. London: Chapman and Hall; 1971

[15] Romberger JA, Hejnowicz Z, Hill JF. The Vascular Cambium. Plant Structure: Function and Development. Berlin, Heidelberg: Springer-Verlag, 1993; pp. 354-385

[16] Roland JC. Early differences between radial walls and tangential walls of actively growing cambial zone. IAWA Bulletin. 1978;**1**:7-10

[17] Cumbie BG. Origin and development of the vascular cambium in *Aeschynomene virginica*. Bulletin of the Torrey Botanical Club. 1984;**111**:42-50

[18] Brown CL. The influence of external pressure on the differentiation of cells and tissues cultured in vitro. In: Zimmermann MH, editor. The Formation of Wood in Forest Trees. New York: Academic Press; 1964; pp. 389-404

[19] Hejnowicz Z. Wzrost intruzywny, podziały poprzeczne i skośne we wrzecionowatych komórkach inicjalnych zranionej miazgi modrzewia. Acta Societatis Botanicorum Poloniae. 1963;**32**:493-503

[20] Hejnowicz Z. Udział wzrostu intruzywnego w procesie zrastanie się miazgi po poprzecznym nacięciu u modrzewia. Acta Societatis Botanicorum Poloniae. 1963;**32**:625-632

[21] Uggla C, Moritz T, Sandberg G, Sundberg B. Auxin as a positional signal in pattern formation in plants. Proceedings of the National Academy of Science. 1996;**93**:9282-9286

[22] Uggla C, Mellerowicz EJ, Sundberg B. Indole-3-acetic acid controls cambial growth in Scots pine by positional signaling. Plant Physiology. 1998;**117**:113-121

[23] Tuominen H, Puech L, Regan S, Fink S, Olsson O, Sundberg B. Cambial-region-specific expression of *the Agrobacterium iaa* genes in transgenic aspen visualized by a linked *uidA* reporter gene. Plant Physiology. 2000;**123**:531-542

[24] Oribe Y, Funada R, Kubo T. Relationships between cambial activity, cell differentiation and the localization of starch in storage tissues around the cambium in locally heated stems of *Abies sachalinensis* (Schmidt) Masters. Trees. 2003;**17**:185-192

[25] Kurczyńska EU. Vessel differentiation in isolated stem segments of *Fraxinus excelsior* L. after treatment with auxin. Acta Societatis Botanicorum Poloniae. 1992;**61**:343-357

[26] Catesson AM, Roland JC. Sequential changes associated with cell wall formation and fusion in the vascular cambium. IAWA Bulletin. 1981;**2**:151-162

[27] Cosgrove DJ. Enzymes and other agents that enhance cell wall extensibility. Annual Review of Plant Physiology and Plant Molecular Biology. 1999;**50**:391-417

[28] Bourquin V, Nishikubo N, Abe H, Brumer H, Denman S, Eklund M, Christiernin M, Teeri TT, Sundberg B, Mellerowicz EJ. Xyloglucan endotransglycosylases have a function during the formation of secondary cell walls of vascular tissues. Plant Cell. 2002;**14**:3073-3088

[29] Du J, Groover A. Transcriptional regulation of secondary growth and wood formation. The International Journal of Plant Biology. 2010;**52**:17-27

[30] Groover AT, Mansfield SD, DiFazio SP, Dupper G, Fontana JR, Millar R, Wang Y. The *Populus* homeobox gene *ARBORKNOX1* reveals overlapping mechanisms regulating the shoot apical meristem and the vascular cambium. Plant Molecular Biology. 2006;**61**:917-932

[31] Chaffey N, Barlow P. The cytoskeleton facilitates a three-dimensional symplasmic continuum in the long-lived ray and axial parenchyma cells of angiosperm trees. Planta. 2001;**213**:811-823

[32] Kurczyńska EU, Hejnowicz Z. Differentiation of circular vessels in isolated segments of *Fraxinus excelsior*. Physiologia Plantarum. 1991;**83**:275-280

[33] Hejnowicz Z, Romberger JA. The common basis of wood grain figures in the systematically changing of cambial fusiform cells. Wood Science and Technology. 1979;**13**:89-96

[34] Hejnowicz Z. The structural mechanism involved in the changes of grain in timber. Acta Societatis Botanicorum Poloniae. 1968;**37**:347-365

[35] Hejnowicz Z. Anatomia i histogeneza roślin naczyniowych. Organy wegetatywne. Warszawa: Wydawnictwo Naukowe PWN; 2002

[36] Aloni R. Ecophysiological implications of vascular differentiation and plant evolution. Trees. 2015;**29**:1-16

[37] Chaffey N, Cholewa E, Regan S, Sundberg B. Secondary xylem development in *Arabidopsis*: A model for wood formation. Physiologia Plantarum. 2002;**114**:594-600

[38] Ragni L, Hardtke CS. Small but thick enough, the Arabidopsis hypocotyl as a model to study secondary growth. Physiologia Plantarum. 2014;**151**:164-171

[39] Sehr EM, Agusti J, Lehner R, Farmer EE, Schwarz M, Greb T. Analysis of secondary growth in the Arabidopsis shoot reveals a positive role of jasmonate signalling in cambium formation. Plant Journal. 2010;**63**:811-822

[40] Agusti J, Lichtenberger R, Schwarz M, Nehlin L, Greb T. Characterization of transcriptome remodeling during cambium formation identifies *MOL1* and *RUL1* as opposing regulators of secondary growth. PLoS Genetics. 2011;**7**:e1001312. DOI: 10.1371/journal.pgen.1001312

[41] Paul-Victor C, Rowe N. Effect of mechanical perturbation on the biomechanics, primary growth and secondary tissue development of inflorescence stems of *Arabidopsis thaliana*. Annals of Botany. 2011;**107**:209-218

[42] Ko J-H, Han K-H, Park S, Yang J. Plant body weight-induced secondary growth in Arabidopsis and its transcription phenotype revealed by whole-transcriptome profiling. Plant Physiology. 2004;**135**:1069-1083

[43] Ko JH, Han KH. Arabidopsis whole-transcriptome profiling defines the features of coordinated regulations that occur during secondary growth. Plant Molecular Biology. 2004;**55**:433-453

[44] Lev-Yadun S, Flaishman MA. The effect of submergence on ontogeny of cambium and secondary xylem and on fiber lignification in inflorescence stems of Arabidopsis. IAWA Journal. 2001;**22**:159-169

[45] Sanchez P, Nehlin L, Greb T. From thin to thick: Major transitions during stem development. Trends in Plant Science. 2012;**17**:113-121

[46] Wabnik K, Kleine-Vehn J, Balla J, Sauer M, Naramoto S, Reinöhl V, Merks RMH, Govaerts W, Friml J. Emergence of tissue polarization from synergy of intracellular and extracellular auxin signaling. Molecular Systems Biology. 2010;**6**:447

[47] Wenzel CL, Schuetz M, Yu Q, Mattsson J. Dynamics of *MONOPTEROS* and PIN-FORMED1 expression during leaf vein pattern formation in *Arabidopsis thaliana*. Plant Journal. 2007;**49**:387-398

[48] Yamaguchi M, Goue N, Igarashi H, Ohtani M, Nakano Y, Mortimer JC, Nishikubo N, Kubo M, Katayama Y, Kakegawa K, Dupree P, Demura T. Vascular-related NAC-DOMAIN6 and vascular-related NAC-DOMAIN7 effectively induce transdifferentiation

in xylem vessel elements under control of an induction system. Plant Physiology. 2010;**153**:906-914

[49] Xu T, Wen M, Nagawa S, Fu Y, Chen J-G, Wu M-J, Perrot-Rechenmann C, Friml J, Jones AM, Yang Z. Cell surface- and Rho GTPase-based auxin signaling controls cellular interdigitation in *Arabidopsis*. Cell. 2010;**143**:99-110

[50] Nieminen KM, Blomster T, Helariutta Y, Mahonen AP. Vascular cambium development. Arabidopsis Book. 2015;**13**:e0177. DOI: 10.1199/tab0177

[51] Mazur E, Kurczyńska EU. Rays, intrusive growth, and storied cambium in the inflorescence stems of *Arabidopsis thaliana* (L.) Heynh. Protoplasma. 2012;**249**:217-220

[52] Chaffey N. Why is there so little research into the cell biology of the secondary vascular system of trees? New Phytologist. 2002;**153**:213-223

[53] Gälweiler L, Guan C, Müller A, Wisman E, Mendgen K, Yephremov A, Palme K. Regulation of polar auxin transport by AtPIN1 in *Arabidopsis* vascular tissue. Science. 1998;**282**:2226-2230

[54] Friml J, Vieten A, Sauer M, Weijers D, Schwarz H, Hamann T, Offringa R, Jürgens G. Efflux-dependent auxin gradients establish the apical-basal axis of *Arabidopsis*. Nature. 2003;**426**:147-153

[55] Benkova E, Michniewicz M, Sauer M, Teichmann T, Seifertova D, Jurgens G, Friml J. Local, efflux-dependent auxin gradient as a common module for plant organ formation. Cell. 2003;**115**:591-602

[56] Petrášek J, Mravec J, Bouchard R, Blakeslee JJ, Abas M, Seifertová D, Wiśniewska J, Tadele Z, Kubeš M, Čovanová M, Dhonukshe P. PIN proteins perform a rate-limiting function in cellular auxin efflux. Science. 2006;**312**:914-918

[57] Demura T, Fukuda H. Novel vascular cell-specific genes whose expression is regulated temporally and spatially during vascular system development. Plant Cell. 1994;**6**:967-981

[58] Fukuda H.. Xylogenesis: Initiation, progression, and cell death. Annual Review of Plant Physiology and Plant Molecular Biology. 1996;**47**:299-325

[59] Gursancky NR, Jouannet V, Grunwald K, Sanchez P, Laaber-Schwarz M, Greb T. MOL1 is required for cambium homeostasis in Arabidopsis. Plant Journal. 2016;**86**:210-220

[60] Uggla C, Magel E, Moritz T, Sundberg B. Function and dynamics of auxin and carbohydrates during earlywood/latewood transition in Scots pine. Plant Physiology. 2001;**125**:2029-2039

[61] Aloni R, Zimmermann MH. Length, width, and pattern of regenerative vessels along strips of vascular tissue. Botanical Gazette. 1984;**145**:50-54

[62] Lev-Yadun S. Wound effects arrest wave phenomena in the secondary xylem of *Rhamnus alaternus* (Rhamnaceae). IAWA Journal. 2001;**22**:295-300

[63] Lev-Yadun S. The distance to which wound effects influence the structure of secondary xylem of decapitated *Pinus pinea*. Journal of Plant Growth Regulation. 2002;**21**:191-196

[64] Warren Wilson J, Pamela M. The position of regenerating cambia—A new hypothesis. New Phytologist. 1961;**60**:63-73

[65] Hejnowicz Z. Tensional stress in the cambium and its developmental significance. American Journal Botany. 1980;**67**:1-5

[66] Savidge RA. In vitro wood formation in "chips" from merchantable stem regions of *Larix laricina*. IAWA Journal. 1993;**14**:3-11

[67] Sachs T, Cohen D. Circular vessels and the control of vascular differentiation in plants. Differentiation. 1982;**21**:22-26

[68] Lev-Yadun S, Aloni R. Vascular differentiation in branch junctions of trees: Circular patterns and functional significance. Trees. 1990;**4**:49-54

[69] Jacobs WP. The role of auxin in differentiation of xylem around a wound. American Journal Botany. 1952;**39**:301-309

[70] Aloni R, Sachs T (1973) The three-dimensional structure of primary phloem systems. Planta. **113**:345-353

[71] Baima S, Nobili F, Sessa G, Lucchetti S, Ruberti I, Morelli G. The expression of the *Athb-8* homeobox gene is restricted to provascular cells in *Arabidopsis thaliana*. Development. 1995;**121**:4171-4182

[72] Baima S, Possenti M, Matteucci A, Wisman E, Altamura MM, Ruberti I, Morelli G. The Arabidopsis ATHB-8 HD-Zip protein acts as a differentiation-promoting transcription factor of the vascular meristems. Plant Physiology. 2001;**126**:643-655

[73] Goldsmith MHM. The polar transport of auxin. Annual Review of Plant Physiology. 1977;**28**:439-478

[74] Rubbery PH, Sheldrake AR.Carrier-mediated auxin transport. Planta. 1974;**118**:101-121

[75] Adamowski M, Friml J. PIN-dependent auxin transport: Action, regulation and evolution. Plant Cell. 2016;**27**:20-32

[76] Morris DA, Thomas J. A microautoradiographic study of auxin transport in the stem of intact pea seedlings (*Pisum sativum* L.). Journal of Experimental Botany. 1978;**29**:147-157

[77] Wetmore RH, Rier JP. Experimental induction of vascular tissues in callus of angiosperms. American Journal Botany. 1963;**50**:418-430

[78] Wodzicki TJ, Zajączkowski S. Effect of auxin on xylem tracheids differentiation in decapitated stems of Pinus silvestris L. and its interaction with some vitamins and growth regulators. Acta Societatis Botanicorum Poloniae. 1978;**43**:129-148

[79] Zajączkowski S. Auxin stimulation of cambial activity in *Pinus silvestris* I. The differential cambial response. Physiologia Plantarum. 1984;**29**:281-289

[80] Vanneste S, Friml J. Auxin: A trigger for change in plant development. Cell. 2009;**136**:1005-1006

[81] Melnyk CW, Schuster C, Leyser O, Meyerowitz EM. A developmental framework for graft formation and vascular reconnection in *Arabidopsis thaliana*. Current Biology. 2015;**25**:1306-1318

[82] Rolland-Lagan A-G, Prusinkiewicz P. Reviewing models of auxin canalization in the context of leaf vein pattern formation in *Arabidopsis*. Plant Journal. 2005;**44**:854-865

[83] Wabnik, K, Robert HS, Smith RS, Friml J. Modeling framework for the establishment of the apical-basal embryonic axis in plants. Current Biology. 2013;**23**:1-6

[84] Friml J, Benková E, Blilou I, Wiśniewska J, Hamann T, Ljung K, Woody S, Sandberg G, Scheres B, Jürgens G, Palme K. AtPIN4 mediates sink-driven auxin gradients and root patterning in *Arabidopsis*. Cell. 2002;**108**:661-673

[85] Friml J, Wiśniewska J, Benková E, Mendgen K, Palme K. Lateral relocation of auxin efflux regulator PIN3 mediates tropism in *Arabidopsis*. Nature. 2002;**415**:806-809

[86] Friml J, Yang X, Michniewicz M, Weijers D, Quint A, Tietz O, Benjamins R, Ouwerkerk PBF, Ljung K, Sandberg G, Hooykaas PJJ, Palme K, Offringa R. A PINOID-dependent binary switch in apical-basal PIN polar targeting directs auxin efflux. Science. 2004;**306**:862-865

[87] Robert HS, Grones P, Stepanova AN, Robles LM, Lokerse AS, Alonso JM, Weijers D, Friml J. Local auxin sources orient the apical-basal axis in Arabidopsis embryos. Current Biology. 2013;**23**:1-7

[88] Grones P, Chen X, Simon S, Kaufmann WA, De Rycke R, Nodzyński T, Zažímalová E, Friml J. Auxin-binding pocket of ABP1 is crucial for its gain-of-function cellular and developmental roles. Journal of Experimental Botany. 2015;**66**:5055-5065

[89] Parry G, Estelle M. Auxin receptors: A new role for F-box proteins. Current Opinion in Cell Biology. 2006;**18**:152-156

[90] Kelley DR, Estelle M. Ubiquitin-mediated control of plant hormone signaling. Plant Physiology. 2012;**160**:47-55

[91] Lavy M, Prigge MJ, Tao S, Shain S, Kuo A, Estelle M. Constitutive auxin response reveals complex interactions between Aux/IAA and ARF proteins in Physcomitrella. eLife. 2016;**5**:1-22. DOI: 10.7554/eLIFE.13325

[92] Paciorek T, Zažímalová E, Ruthardt N, Petrášek J, Stierhof Y-D, Kleine-Vehn J, Morris DA, Emans N, Jürgens G, Geldner N, Friml J. Auxin inhibits endocytosis and promotes its own efflux from cells. Nature. 2005;**435**:1251-1256

[93] Robert S, Kleine-Vehn J, Barbez E, Sauer M, Paciorek T, Baster P, Vanneste S, Zhang J. ABP1 mediates auxin inhibition of clathrin-dependent endocytosis in *Arabidopsis*. Cell. 2010;**143**:111-121

[94] Gao Y, Zhang Y, Zhang D, Dai X, Estelle M, Zhao Y. Auxin binding protein 1 (ABP1) is not required for either auxin signaling or Arabidopsis development. Proceedings of the National Academy of Sciences of the United States of America. 2015;**112**:2275-2280

[95] Michalko J, Glanc M, Perrot-Rechenmann C, Friml J. Strong morphological defects in conditional *Arabidopsis abp1* knock-down mutants generated in absence of functional ABP1 protein. F1000Research. 2016;**5**:86

[96] Zhao C, Johnson BJ, Kositsup B, Beers EP. Exploiting secondary growth in Arabidopsis. Construction of xylem and bark cDNA libraries and cloning of three xylem endopeptidases. Plant Physiology. 2000;**123**:1185-1196

[97] Etchells JP, Provost CM, Mishra L, Turner SR. WOX4 and WOX14 act downstream of the PXY receptor kinase to regulate plant vascular proliferation independently of any role in vascular organization. Development. 2013;**140**:2224-2234

[98] Chaffey N, Barlow P, Barnett J. Cortical microtubules rearrange during differentiation of vascular cambial derivatives, microfilaments do not. Trees. 1997;**11**:333-341

[99] Hagen G, Guilfoyle TJ. Rapid induction of selective transcription by auxin. Molecular and Cellular Biology. 1985;**5**:1197-1203

[100] Mitsuda N, Iwase A, Yamamoto H, Yoshida M, Seki M, Shinozaki K, Ohme-Takagi M. NAC transcription factors, NST1 and NST3, are key regulators of the formation of secondary walls in woody tissue of *Arabidopsis*. Plant Cell. 2007;**19**:270-280

[101] Zhong R, Richardson EA, Ye ZH. Two NAC domain transcription factors, SND1 and NST1, function redundantly in regulation of secondary wall synthesis in fibers of *Arabidopsis*. Planta. 2007;**225**:1603-1611

[102] Altamura MM, Possenti M, Matteucci A, Baima S, Ruberti I, Morelli G. Development of the vascular system in the inflorescence stem of *Arabidopsis*. New Phytologist. 2001;**151**:381-389

[103] Lev-Yadun S. Fibres and fibre-sclereids in wild-type *Arabidopsis thaliana*. Annals of Botany. 1997;**80**:125-129

Section 5

Fighting the Climate Changes

Chapter 8

Climate Smart Agriculture: An Option for Changing Climatic Situation

Mona Nagargade, Vishal Tyagi and
Manoj Kumar Singh

Additional information is available at the end of the chapter

http://dx.doi.org/10.5772/intechopen.69971

Abstract

World population is increasing day by day and at the same time agriculture is threatened due to natural resource degradation and climate change. Production stability, agricultural productivity, income and food security is negatively affected by changing climate. Therefore, agriculture must change according to present situation for meeting the need of food security and also withstanding under changing climatic situation. Projected estimates based on food consumption pattern and population growth show that agriculture production will require enhancing by 65% to meet the need of burgeoning population by 2050. Agriculture is a prominent source as well as a sink of greenhouse gases (GHGs). So there is a need to modify agricultural practices in a more sustainable way to overcome these problems. Developing climate-resilient agriculture is thus crucial to achieving future food security and climate change goals. It helps the agricultural system to resist damage and recover quickly by adaptation and mitigation strategies. Mitigation strategies reduce the contribution of agriculture system to greenhouse gas emission, and adaptation strategies provide agriculture production under changing scenarios. This chapter explains different mitigation and adaptation strategies, including farming practices and engineering approaches.

Keywords: adaptation, climate change, climate smart agriculture, mitigation, sustainable development

1. Introduction

Climate change is emerging as a major threat on agriculture, livelihood and food security of millions of people in many places of the world [1]. At the same time, many current farming practices damage the environment and are a major source of major greenhouse gases (GHGs), that is, carbon dioxide (CO_2), methane and nitrous oxide. Annual anthropogenic greenhouse gas (GHG) emissions that are classified in Intergovernmental Panel on Climate Change (IPCC) reports as originating in 'agriculture, forestry and other land use' (AFOLU) are caused mainly by deforestation, rice cultivation practices, livestock production, soil and nutrient management. It has been estimated that AFOLU contributes to 21% of total global emission. By Food and Agricultural Organization (FAO) estimates, emissions from AFOLU stood at 10.6 Gigatonnes (Gt) of carbon dioxide equivalent in the year 2014. Large deforestation and land degradation have also reduced the carbon sequestration capacity from the atmosphere. Forestry, agriculture and land use accounts for 49 and 30% of total emission of carbon dioxide and methane, respectively. The share of nitrous oxide in total AFOLU emissions is small, but accounts for as much as 75% of global anthropogenic emissions of the gas. Forests play important role in climate mitigation by removing a large amount of GHG from the atmosphere. However, the average contribution of forests in carbon sequestration has fallen from 2.8 Gt annually in the 1990s to 2.3 Gt in the 2000s and is estimated at 1.8 Gt in 2014 [2].

Worldwide, more than a billion of farmers and their families face a great challenge of climate change because agriculture sector is the most vulnerable sector for climate change. Thus, their lives and livelihoods are directly affected by climate change, so there is an urgent need to implement many of the solutions to overcome this problem. Despite the attention paid to agricultural development and food security over the past decades, there are still about 800 million undernourished and 1 billion malnourished people in the world. At the same time, more than 1.4 billion adults are overweight and one-third of all food produced is wasted. Before 2050, the global population is expected to swell to more than 9.7 billion people [3]. At the same time, global food consumption trends are drastically changing, for example, increasing demand for meat-rich diets. If the current trends in consumption patterns and food waste continue, it is estimated that we will require 60% more food production by 2050 [4]. The International Food Policy Research Institute (IFPRI) estimated that by 2050 about 50 million more people could be at a risk of undernourishment because of climate change.

2. Impact of climate change on agriculture

Climate change is very likely to affect food security at the global, regional and local level. Climate change can disrupt food availability, reduce access to food and affect food quality [5]. Many advance technologies, such as improved crop varieties, pest control methods, genetically modified organisms and irrigation systems, are widely adopted for crop improvement. By 1880, 1.5°C of Earth's temperature had been increased [6]. It is projected that throughout the twenty-first century, India will experience warming above global level. It is also expected that India will experience more seasonal variation in temperature with more warming in the winters than in the summers. In recent years, extreme heat waves across India have extended together with warmer night temperatures and hotter days, and this trend is expected to continue. This is predicted that the average temperature will raise up to 2.33–4.78°C with a doubling in CO_2 concentrations at the end of this century. These heat waves will lead to increased variability in summer monsoon precipitation, which will result in drastic effects on the agriculture sector in India [7]. Due to rainfall variability, changes in the frequency and severity of droughts and floods could pose challenges for farmers and ranchers and will ultimately threaten global food security [8]. Meanwhile, the rise in sea level and warming of sea water are likely to cause the habitat ranges of many fish species to shift, which could disrupt ecosystems. Overall, climate change would adversely affect over crops, animals and marine life. Along with effects of climate change on agriculture, other evolving factors that affect agricultural production, such as changes in farming practices and technology, need to be addressed.

Higher CO_2 levels can affect crop yields. The elevated CO_2 levels can increase plant growth particularly in plants with C_3 photosynthetic mechanism. Increased CO_2level also reduces the quality of produce by reducing protein, nitrogen and minerals content in most plant species, including wheat, soybeans and rice; this will threaten human health. Climate change also increased pest pressures and reductions in the efficacy of pesticides; thus more pesticides will be required in future which also threatened human health [8]. Reduced grain and forage quality can reduce the ability of pasture and rangeland to support grazing livestock [9]. Climate change also affects weeds, pests and fungi infestation due to an increase in host susceptibility. The ranges and distribution of weeds and pests are likely to increase with climate change. This could cause new problems for farmers' crops previously unexposed to these species. Both historical and future estimated impacts of climate change on cereal crop yields in different locations indicate that the yield loss can be up to −35% for rice, −20% for wheat, −50% for sorghum, −60% for maize and −13% for barley depending on the location, future climate scenarios and projected year [10].

Heat stress also affects animals both directly and indirectly. Over time, heat stress can increase vulnerability to disease, and reduce fertility and milk production. Climate change also affects rainfall pattern and distribution; due to this, many areas are facing a problem of drought stress which may threaten pasture and feed supplies. Drought reduces both quantity and quality of forage available to grazing livestock. Some areas could experience longer, more intense droughts, resulting from higher summer temperatures and reduced precipitation. Changes in crop production due to drought could also become a problem for animals that rely on grain. Climate change may increase the prevalence of parasites and diseases that affect livestock. More variability in weather pattern could provide the chance of insect, pest infestation in crop plant [11]. Pasture productivity is increased by elevated CO_2but its quality may be reduced due to high lignin concentration, low protein and digestibility. However, the quality of some of the forages found in pasturelands decreases with higher CO_2. As a result, the overall performance of cattle would be affected and require eating more to get the same nutritional benefits.

Fisheries and aquaculture, which provide at least 50% of animal protein to millions of people in low-income countries, are already under multiple stresses, including overfishing, habitat loss and water pollution [12]. Changes in temperature and seasons can also affect the timing of reproduction and migration fishes and other aquatic life. Lifecycle of aquatic animals is controlled by the temperature of aquatic system and the change in the seasons. For example, in Northwest (United States), the lifecycle of salmon may be affected by an increase in the temperature of water and this could also increase the likelihood of disease. These effects in combination of other climate impacts are projected to lead to large declines in salmon populations [13]. The increased intensity and frequency of storms, hurricanes and cyclones will adversely affect coastal fisheries, aquaculture and mangroves.

Forests provide employment and livelihood for more than 100 million people of the world's rural poor. Forests are home to more than 80% of the world's earth biodiversity and provide ecosystem services like food, fuel, timber, medicines and shelter. Aberrant weather limits the capacity of forests to produce goods and services and also affect the people who depend on them directly or indirectly. On the other hand, temperate forest region communities will be benefitted from elevated CO_2 and temperature. While due to changes in occurrence, distribution and frequency of precipitation, this will more likely affect natural biodiversity and also adversely affect crop growth, productivity and ultimately yield [14].

India and China, the most densely populated countries of the world, to keep pace with the growing population and to sustain world food security will require maintaining at least 4–5% annual growth rate in agriculture. India supports about 17% of human and 11% of livestock population of the world just on 4.2% of water resources and 2.8% land [15]. As per recent estimates, India will need to produce about 281 million tonne (mt) food grains, 53.7 mt oilseeds, 22 mt pulses, 127 mt vegetables and 86 mt fruits by 2020–2021 [15]. In India, the average food consumption at present is 550 g per capita per day, whereas in China and USA are 980 and 2850 g, respectively [16, 17]. To meet the demand for food from this increased population, the country's farmers need to produce 50% more grain by 2020 [7]. To enhance food production is under changing climatic conditions like aberrant weather, rising CO_2 level, rising temperature and rising sea level, and we require the reorientation of agriculture from current practices to more sustainable and environmental friendly practices. In this context, scientists are more focused on climate smart or resilience agriculture (CSA). Climate resilience agriculture helps to improve food security under changing climatic situation while also reducing food waste globally [18] and minimizing danger on natural resources.

2.1. Definition

More specifically for managing agriculture for food security under the changing realities of climate, FAO has developed the 'Climate Smart Agriculture' (CSA) approach, which it presented in 2010 at The Hague Conference on Agriculture, Food Security and Climate Change [19]. United Nations (FAO) defines CSA as 'Agriculture that sustainably increases productivity, enhances resilience (adaptation), reduces/removes GHGs (mitigation) where possible, and enhances achievement of national food security and development goals'. The country will have to produce more quality food with diminishing natural resource base and changing climate. It includes an in-built property in the system for the recognition of a threat that needs to be responded to and also the degree of effectiveness of the responses.

2.2. Climate smart strategies

Many agricultural technologies and practices such as minimum tillage, different methods of crop establishment, nutrient and irrigation management and residue management can improve crop yields: nutrient and water use efficiency and reduced greenhouse gas (GHG) emissions from agricultural activities [20]. Similarly, the use of improved seeds, rainwater harvesting (RH), Information and Communication Technologies (ICTs)-based agro-advisories and crop/livestock insurances can also help farmers to reduce the impact of climate change and variability [21] In general, the CSA options integrate innovative and traditional technologies, practices and services that are relevant for particular location and reduce the effect of climate change and provide the opportunities to stand such changing scenario. Adaptation and mitigation are complementary strategies for reducing and managing the risks of climate change. Substantial reduction in GHG emission over the next few decades can reduce the occurrence of climatic variability in the twenty-first century and beyond, increase prospects for effective adaptation, reduce the costs and challenges of mitigation in the longer term and contribute to climate-resilient pathways for sustainable agriculture.

2.3. Mitigation strategies

Mitigation practices help to reduce the emission of greenhouse gases from agriculture system because elevated greenhouse gases ultimately lead to climate change.

2.3.1. Mitigation through farming practices

A change in the system of cultivation, cultural practices, nutrient management and water management helps to reduce the emission of greenhouse gases (**Table 1**).

Conservation practices	GHG objectives	Additional benefits
Crops		
Conservation tillage and traffic control	Capture carbon and reduce emission	Improves soil, water and air quality. Soil erosion, compaction and fuel use are also reduced.
Integrated and site-specific nutrient management	Capture carbon and reduce emission	Improves water and air quality and also saves time, labour and money.
Crop diversification and crop rotation	Carbon sequestration	Improves soil and water quality and also reduces emission and water requirement and provides food security.
Animals		
Manure management	Reduces emission	Improves soil quality, crop yield, on-farm sources of biogas fuel and possibly electricity for large operations, provides nutrients for crops.
Rangeland and pasture management through rotational grazing and improved forage	Sequestration, emission reduction	Reduces water requirement, helps in withstand drought. Increases long-term grassland productivity.
Feed management	Emission reduction	Reduces quantity of nutrients. Improves water quality. More efficient use of feed.

Source: NRCS. http://soils.usda.gov/survey/global-climate-change.html.

Table 1. Agricultural practices and climate change mitigation.

2.3.1.1. Methane emission

Methane is about 25 times more effective as a heat-trapping gas than CO_2. The main sources of methane are wetlands, organic decay, termites, natural gas and oil extraction, biomass burning, rice cultivation, enteric fermentation and refuse landfills. CH_4 emissions can be controlled by the following practices:

- Controlling production, oxidation and transport of CH_4 from paddy field by alteration in water management, particularly promoting mid-season aeration, by short-term drainage, direct-seeded rice (DSR) cultivation. The IPCC [6] has estimated that rice cultivation is a major contribution to global warming. Rice cultivation contributes 23% of total greenhouse gas emission from agriculture sector. Under continuous flooding, dry-DSR emitted 24–79% of CH_4 as compared with 8–22% in wet-DSR, whereas, under intermittent irrigation, the reduction ranged from 43 to 75% in dry-DSR compared with conventional tilled transplanted rice (CT-TPR). However, reduction in CH_4 emissions increased further when DSR was combined with mid-season drainage or intermittent irrigation, compared with flooded CT-TPR. The reason for low CH_4 emissions from dry-DSR is aerobic conditions, especially during the early growth stage [22]. Methane emission starts at a redox potential of soil below −150 mV and is stimulated at less than −200 mV [23].
- Improved organic matter management by promoting aerobic degradation through incorporating or composting it into soil during off-season-drained period.
- Use of rice cultivars with few unproductive tillers, high root oxidative activity and high harvest index.
- Improved management of livestock diet through the use of improved feed additives, substitution of low digestibility feeds with high digestibility ones.

2.3.1.2. Nitrous oxide

As a greenhouse, nitrous oxide is 298 times more effective than CO_2. Forests, grasslands, oceans, soils, nitrogenous fertilizers and burning of biomass and fossil fuels are the major sources of nitrous oxide. Site-specific nutrient management (SSNM), integrated nutrient management (INM), use of slow-release nitrogen fertilizers and nitrification inhibitors, and placement of fertilizer in reduced zone helps to reduce emission of NO_2.

2.3.1.3. Carbon dioxide

The main sources of carbon dioxide emission are decay of organic matter, forest fires and eruption of volcanoes, burning of fossil fuels, deforestation and land-use changes. CO_2 emission mitigates by carbon sequestration, soil management practices and afforestation. Carbon sequestration is one of the best strategies to mitigate CO_2 emission, which is the removal of atmospheric CO_2into long-lived stable form.

- Carbon sequestration in soil through manipulation of soil moisture and temperature and restoration of soil carbon on degraded lands. Trees, plants and crops absorb CO_2 naturally through photosynthesis and stored as carbon in biomass in tree trunks, branches, foliage and roots and soils [24]. Forests and stable grasslands are good sink of carbon because they can store large amounts of carbon in their vegetation and root systems in long-life stable form as compared to field crops. Soils are the largest terrestrial sink for carbon on the planet.

- Soil management practices such as reduced tillage and manuring residue incorporation. Improving soil biodiversity, micro-aggregation and mulching.

2.3.2. Mitigation through transgenic crops

Agriculture contributes significantly to greenhouse gas (GHG) emissions. As indicated by researchers, there is a need to give priorities to develop new varieties of crops cultivars that can reduce GHG emissions in farming systems [25]. In this regard, GM crops have been contributing to lower GHG emissions through reducing fuel use, due both to less pesticide applications and increasing the area grown under conservation agriculture (CA), which involves practices such as 'no-till' or 'reduced-till'. It is reported that, in 2012, GM crops were grown roughly on 12% of the world's arable land, which reduced over 26.7 billion kg of carbon dioxide (CO_2) or the equivalent of removing nearly 12 million cars from circulation with a total reduction [26]. Genetically modified crops such as Roundup Ready™ modify the plants through genetic engineering so that they can take more carbon from atmosphere and convert it to oxygen [27].

Crops are bred for N-use efficiency (NUE) because this trait is a key factor for reducing N fertilizer pollution, improving yields in N-limited environments and reducing fertilizer costs. There are various genetic engineering activities which improve NUE in crops [28]. The alanine aminotransferase gene from barley, which catalyses a reversible transamination reaction in the N-assimilation pathway and enhances NUE, appears to be a target for plant breeding option. Keeping N in ammonium form will affect how N remains available for crop uptake and will improve N-recovery, thus reducing groundwater and the atmosphere pollution. There are genes in tropical grasses such as *Brachiaria humidicola* and in the wheat wild relative *Leymus racemosus* that inhibit soil nitrification by releasing inhibitory compounds from roots and suppressing nitrosomonas bacteria [29]. Engineering nitrogen fixation for self-fertilizing crops can also help to reduce the use of synthetic fertilizer. The ultimate goal is to transfer the molecular machinery for nitrogen fixation into non-N_2-fixing plants so that they will fix atmospheric N_2 [30]. The process is very complex and challenging because N_2-fixing nitrogenase enzyme is adversely affected by oxygen produced during photosynthesis. But agriculture would be revolutionized if plants can be engineered to fix their own nitrogen; this would free agriculture from synthetic nitrogenous fertilizers and significantly decouple it from the fossil fuel industry [31].

About one-fifth of global CH_4emissions are from enteric fermentation in ruminant animals. Apart from various rumen manipulations and emission control strategies, genetic engineering is a promising tool to reduce these emissions. The most practical and rapid mitigation procedure may be to reduce the per cow CH_4 emission through animal breeding and genetic selection for feed efficiency as it is permanent and cumulative [32]. Other options like manipulation in diet composition, supplementation of feed additives and selection of forage plants of high quality for breeding provide solution for reduced CH_4 emission from enteric fermentation.

2.4. Adaptation strategies

Adaptation is a key factor that will shape the future severity of climate change impacts on food production. To deal with the impact of climate change, the potential adaptation strategies are as follows:

2.4.1. Adaptation through transgenic approaches

Transgenic approaches are one of the many tools available for modern plant improvement programmes. This approach involves manipulation of genetic material and the fusion of cells beyond normal breeding barriers. The most evident example of genetic engineering is 'transgenic' technology which involves the insertion or deletion of genes and creating genetically modified organisms (GMOs). In genetic engineering or genetic transformation, the genetic material is modified by artificial means. It involves isolation and cutting of a gene at a precise location by using specific enzymes. Identification and transfer of stress-tolerance gene into plant provide improved productivity and adaptation to abiotic stresses. Hence, the transformation of major crop species with genes from any biological source (plant, animal and microbial) is the most powerful tool for molecular plant breeding. Transgenic plant can be used as sources of new cultivars and they are also extremely useful as proof-of-concept tools to dissect and characterize the activity and interplay of gene networks for abiotic stress resistance. Transcription factor (TF), and genes involved in adaptation to extreme temperature, drought or salinity in crops are [33] as follows:

1. Involved in sense, percept and transduction of signal.
2. Mechanisms of stress response for adaptation and abscisic acid (ABA) biosynthesis for enhancement for drought adaptation.
3. Genes involved in transportation, detoxification and signal transduction as well as TF tolerance to salinity.
4. Some associated to reactive oxygen species (ROS), modification of membrane and production of chaperon, late embryogenesis abundance (LEA) proteins, osmoprotectants/ compitable solutes and TF are continued in crop genetic engineering for temperature extremes.

2.4.1.1. Heat

Plant responses to high temperature vary with the degree and duration of high temperature and the plant type. Plants have various mechanisms to cope with high temperatures, for example, by maintaining membrane stability, or by ion transporters, proteins, osmoprotectants, antioxidants and other factors involved in signalling cascades and transcriptional control [34]. The activity of the enzymes CAT, glutathione S-transferase (GST) and APX was more enhanced in the cultivar and showed better tolerance to heat stress and projection against ROS production. Upregulation of different genes helps the plant to withstand the stress conditions and also adaptation to stress [35]. Stress plants receive the internal and external signals by different interrelated or independent pathways which are used to control various responses for its tolerance development [36]. Some stress-associated genes such as ROB5—a stress-inducible gene isolated from bromegrass—enhanced the performance of transgenic canola and potato at high temperatures [37]. Likewise, researchers introduced hsp101—a heat shock protein gene from Arabidopsis—in basmati rice. This transgenic rice had a better growth in the recovery phase after suffering heat stress [38].

2.4.1.2. Drought

Several drought-tolerant transgenic plants, including rice, tomato, soybean, maize, barley and Arabidopsis have been developed. Such genetically engineered plants have generally been developed using gene-encoding proteins that control drought-regulatory networks. Stress-signalling networks in drought responses are made of intercellular communication systems, intracellular signalling systems and transcriptional-regulatory complexes [39]. Some important proteins like protein kinases, receptor-like kinases, transcription factors, some enzymes related to osmoprotectant or plant hormone synthesis, and other functional and regulatory proteins [40] play an important role in plant defence during drought stress. NAC, AP2/ERF, bZIP and MYC orchestrate regulatory networks are major transcription factor families of plants, which underlie drought stress tolerance [41]. Enhanced adaptation to drought could be provided directly by metabolites such as trehalose, mannitol, glycinebetaine or indirectly through regulation of gene expression through TF and kinases in signal transduction [42]. The Arabidopsis' HARDY (HRD) gene improved WUE by enhancing photosynthetic assimilation and reducing transpiration [43]. Further research by a multinational seed company [44] revealed that the encoding gene cold shock protein B (CspB) from *Bacillus subtilis*, a soil bacterium, allows the transgenic maize plant to react more quickly to drought, slowing its growth and conserving water, thereby making water available for key plant functions after the onset of drought stress. The DroughtGard™ hybrid maize was bred and released for farming in the USA in 2013. Under stress, a DroughtGard™ hybrid used 261 mm of water from the soil while the control used 338 mm of water from the soil. Under drought stress, a maize plant begins producing sugars such as trehalose, which can be broken by the glycoside hydrolase enzyme trehalase. When this sugar is not destroyed, the plant shows enhanced adaptation to drought stress.

2.4.1.3. Salinity

The damaging effects of salt accumulation in agricultural soils have severely affected agricultural productivity in large swathes of arable land throughout the world [45]. Salt-affected land accounts for more than 6% of the world's total land area. Climate change also causes salinity due to effect on soil water and increase in temperature. Therefore, there is an urgent need to develop salt-tolerant varieties of crop through conventional breeding and transgenic approaches. Genes enhance tolerance in plant against salinity by employing several mechanisms such as by limiting uptake rate of salts from soil and further transport of salts inside plant system, regulate leaf development and the onset of plant senescence and adjust the ionic and osmotic balance of cells in roots and shoots. Salt tolerance mechanism varies with the plant species; however, the capacity to maintain low cytosolic Na^+ is thought to be one of the key determinants of plant salt tolerance [46]. Transgenic plants accumulated more K^+, Ca^{2+} and Mg^{2+} and less Na^+ in their shoots compared with non-transformed controls. The most promising genes for the genetic engineering of salinity tolerance in crops, as noted by scientist [47], are related to ion transporters and their regulators, as well as the C-repeat-binding factor [48]. The successful use of transporter genes has been observed in several plants. He et al. developed transgenic cotton plants with AtNHX1 expression and revealed that plants produced more fibre content and generated more biomass under salt stress in a greenhouse condition [49]. Increased fibre yield was due to superior photosynthetic performance and higher nitrogen assimilation rates in the transgenic plants compared with wild type. Moghaieb et al. developed transgenic tomato plants by isolating gene (which produce ectoine, a compatible osmolite) from bacteria and inserting in tomato plant [50]. He reported that transgenic plant produced a high amount of ectoine under saline conditions, which led to increased sink capacity for photosynthate in root and also enhanced water uptake, and increased the photosynthetic rate under salt stress through enhancing cell membrane stability [51].

2.4.2. Developing climate-ready crops

Development of new crop varieties with higher yield potential and resistance to multiple stresses (biotic and abiotic) will be the key to maintain yield stability. Several drought tolerance varieties have been released in South and Southeast Asia. For example, in 2010 the variety Sahbhagi Dhan, released and notified in India, showed a consistently good performance under transplanted low-land conditions and rain-fed direct-seeded upland [52]. Singh reported 108% higher yields with Sahbhagi Dhan compared with popular local varieties under water-scare situation [53]. Sahbhagi Dhan needed one to two irrigations only to yield 4.7 t ha^{-1}. Likewise, Sahbhagi dhan, IR64-Drought1, is another climate-ready rice [54]. Nitrogen use efficiency may be reduced under the climate change scenarios because of heavy precipitation and high temperature which increase leaching and volatilization losses, respectively. But at the same time, elevated CO_2 concentration causes plant growth and development which leads to increased demand for nitrogen. So, there is a need to develop the improved root system of plant for nutrient and water absorption. Genetic engineering is the best option for pooling all the desirable traits in a plant to get the ideal plant type for adverse climatic situation.

2.4.3. Crop diversification

Major shift in terms of diversification of agriculture into crops, commodities, enterprises and cropping/farming systems is called upon to revert the process of degradation of natural resources, rejuvenations of waste lands and also to make agriculture a profitable business. Diversified agricultural systems may be a productive way to build resilience into agricultural systems. Crop diversification helps farmer against aberrant weather conditions like early season drought, late season drought and dry spell during crop growth season. Intercropping of soybean + pigeonpea (4:2), pearlmillet + pigeonpea (3:3), pigeonpea + green gram (1:2) and cotton + green gram (1:1) are more economic than monocropping [55]. Due to growing income and urbanization, demand for high-value food products, such as fruits, vegetables, dairy, meat, eggs and fish, is increasing. This is reducing the demand for traditional rice and wheat. Diversification of rice-wheat system

(RWS) towards high-value commodities will increase income and result in reduced water, fertilizer and other resource uses. Pulses have a positive impact on soil quality because they help fix nitrogen in the soil. Nitrogen-fixing pulse crops have a lower carbon and water footprint as compared to other crops. For instance, the water footprints to produce a kilogram of beef, pork, chicken and soybeans are 43, 18, 11 and 5 times higher than the water footprint of pulses. In fact, one study showed that 1 kg of legume only emits 0.5 kg in CO_2 equivalent, whereas 1 kg of beef produces 9.5 kg in CO_2 equivalent. So the inclusion of pulses in crop rotation and intercropping system shows better productivity, sustainability and environmental safety benefits [56].

2.4.4. Alteration in land-use pattern

Change in location of crop and livestock, adjustment in cropping pattern, planting time and methods, fertilizer and pesticide use pattern, and other management practices help to reduce the risk of climate. Alternate land-use management practices also reduce disease and pest outbreak and provide remunerative production under aberrant weather situation.

2.4.5. Changing cropping season

Yield instability is reduced by changing planting time which reduce the impact of temperature increase-induced spikelet sterility and also avoid flowering period to coincide with the hottest period. Crop calendar provides the information about crop location and cropping pattern based on weather pattern which helps the farmer for growing crop according to the occurrence of weather events. Cropping systems may have to be altered to the growth of suitable cultivars, increasing crop intensities (i.e. the number of successive crop produced per unit area per year) or planting different types of crops. Farmers will have to adapt to changing situation by changing crops. Singh et al. showed that the lentil varieties Pusa Vaibhav and Mallika matched well with Sahbhagi Dhan in drought-prone area [57]. Both lentil varieties are of medium duration and suitable for rice-fallow areas of eastern India.

2.4.6. Conservation agriculture

Conservation agriculture (CA) is defined as resource-saving agriculture crop production that strives to achieve acceptable profit together with high and sustained production level while concurrently conserving the environment. It also enhances natural biological process above and below the ground. CA is characterized by three linked principles, namely

I. *Minimum mechanical soil disturbance:* Excessive tillage of agricultural soils may result in short-term increases in fertility, but will degrade soils in the medium term. Structural degradation, loss of organic matter, erosion and falling biodiversity are all to be expected.

II. *Permanent organic soil cover*: Keeping the soil covered and planting through the mulch will protect the soil and improve the growing environment for the crop.

III. *Diversification of crop species grown in sequences/association*: Diversification of crop or cropping system helps farmers in terms of risk minimization. It provides opportunity to crop for efficient utilization of natural resources and also maintain soil fertility.

Conservation agriculture-based resource conservation technologies (RCTs) under precision laser land levelling are helpful to improve water productivity [58]. Retention of carbon (C) in arable soils has been considered as a potential mechanism to mitigate soil degradation and to sustain crop productivity. The mean annual input of organic biomass/residues to soil from all crops varied with aboveground yield responses of the crops and treatment types. The total estimated C input (12.1 Mg C ha^{-1} in 3 years) under mungbean residue + direct-seeded rice (DSR) followed by zero-tilled wheat (ZTW) with rice residue (RR) retention and zero-tilled relay summer mungbean (MBR + DSR-ZTW + RR-ZTMB) treated plots was 117 and 127% higher than DSR-ZTW and transplanted rice followed by conventional tilled wheat (TPR-CTW) treatments, respectively [59]. Jat et al. also reported that various components of CA increase crop growth and productivity [60].

2.4.7. *Efficient utilization of resources*

The resource-efficient technologies comprises those technologies which improve resource use efficiency and provide immediate economic benefits like conservation of natural resources (water, soil, biodiversity and climate), reduce production cost, reduce environmental pollution and ultimately increase yield and income of small and marginal farmers. Resource-conserving practices like zero tillage can allow farmers to sow wheat sooner after rice harvest, so the crop heads and fills the grain before the onset of premonsoon hot weather. The resource conservation technologies (RCTs) in rice-wheat system (RWS) also have pronounced effects on mitigation of greenhouse gas emission and adaptation to climate change. Grace et al. reported that the emissions of GHGs from RWS in Indo-Gangetic plains (IGPs) have a global warming potential (GWP) of 13–26 Mg CO_2 $ha^{-1}yr^{-1}$[61]. These approaches of crop management should be coupled with the measures of crop improvement for wider adaptation to climate change. Rice-wheat cropping systems (RWCS) of the IGP of India are tillage, water and energy intensive and an important source of greenhouse gas (GHG) emission. ZTW – DSR and zero-till wheat + surface application of rice residue – Direct-seeded rice showed the lowest global warming potential (GWP) and GHG intensity. Adoption of these systems in the Indian IGP can reduce GWP of the conventional RWCS (CTW-TPR) by 44–47% without any significant reduction in the system yield. This was mainly due to prolonged aerobic condition under DSR which led to low CH_4 emission (82.3–87.2%) as compared to TPR. However, frequent wetting and drying led to higher denitrification emissions of N_2O (60–70%) in DSR system. Neem oil-coated urea was found to be effective in reducing N_2O emission from ZTW (17.8–20.5%) leading to lower GWP as compared to CTW. Application of rice residue in ZTW treatment also reduced N_2O emission (11–12.8%) [62].

2.4.8. *Integrated nutrient management*

Integrated nutrient management (INM) system or integrated plant nutrient supply (IPNS) system is a practice which aims at achieving a harmony by efficient and judicial use of chemical fertilizers in conjunction with organic manures, use of well-decomposed crop residues, green manures, recyclable waste, compost including vermicompost, using legumes in cropping systems, use of bio-fertilizers and other locally available nutrient sources for sustaining soil health and amelioration of environment as well as enhancing crop productivity on long-term basis [63]. Minimum soil disturbance of reduce tillage, integrated and judicious use of different nutrient sources may enhance soil organic carbon marginally and the carbon sequestration rate varied from 62 to 186 kg ha^{-1} yr^{-1}. Based on the study, it can be recommended that the substitution of 50% of the recommended N with organic source increases crop yields and soil carbon in semi-arid rain-fed systems of India [64].

2.4.9. Site-specific nutrient management

Site-specific nutrient management (SSNM) is a plant-based approach for managing the nutrient requirements of crop. It provides principles and tools for supplying nutrients as and when needed for plant to achieve high yields while optimizing the use of nutrients from indigenous sources. Applying the right nutrient source, at the right rate, at the right time, in the right place is essential to nutrient stewardship. LCC-based urea application can reduce GWP of a rice-wheat system by 10.5% in LCC≤4 treatment as compared to blanket application [65].

2.4.10. Relocation of crops

Climatic variabilities such as increased temperature, CO_2 level, drought and floods would affect the production of crops. However, the impact will be varied across crops and regions. There is a need to identify the regions and crops that are more sensitive to climate changes/ variability and relocate them in more suitable areas.

2.4.11. Harnessing indigenous technical knowledge of farmers

There is a wealth of knowledge on the range of measures that can help in developing technologies to overcome climate vulnerabilities. There is a need to harness the indigenous technical knowledge and fine-tune them to suit the modern situation. Ecological-based traditional knowledge could provide insights and viable options for adaptive measures.

2.4.12. Integrated Farming System (IFS)

Monocropping in flood- and drought-prone area is risky practices for farmers. Dependence on single enterprises not only increases the risk of crop failure but also leads to food, income and environmental insecurity especially in rain-fed area. Integrated farming system (IFS) modules minimize risk from a single enterprise in the face of natural calamities, and diversified enterprises bring in the much needed year round income to farmers in monocropped paddy-growing areas and improve their livelihoods and resilience to extreme weather events. Integrated farming system is defined as the integration of different interrelated, interacting and interdependent farm enterprises which are suited to agroclimatic condition socioeconomic situation of the farmers. Integrated fish-duck farming and Rice-fish-poultry farming have been developed for small and marginal farmers [55].

2.4.13. Integrated pest management

Several factors mainly climatic factors such as variability in rainfall and changes in temperature would affect the incidence of pest, disease and host susceptibility of major crops, because

climate is continuously changing which will potentially influence the pest/weed-host relationship by affecting the pest/weed population, the host population and the pest/weed-host interactions. To adopt in this situation, some of the potential strategies are as follows:
Develop diseases and pest resistance cultivars,
Adopt integrated pest management with more emphasis on biological control,
Improve forecasting of pest using recent tools and techniques such as simulation modelling,
Develop location-specific crops, cultivars and alternative production techniques that are resistant to infestations and other risks.

2.4.14. Better weather forecasting and crop insurance schemes

Weather forecasting at different spatial and temporal scales would be significant tool for adaption in agriculture under future climate change scenario. Information and Communication Technologies can also play a great role to disseminate the information [66]. Weather forecasting and early warning systems will be very useful in minimizing risks associated with climatic adversaries. Information and Communication Technologies could greatly help the administrators and researchers in developing contingency crop plans and also reduce the risk of aberrant weather. Efficacious crop insurance schemes should be evolved to help the farmers in reducing the risk of crop failure due to these events. Low-cost access to financial services could be a boon for vulnerable farmers. Micro-finance has been a success among rural poor, including women. Rainwater harvesting, drip irrigation, laser land levelling, furrow-irrigated raised bed planting, drainage management, cover crop method, site-specific nutrient management, green manuring, integrated nutrient management, intercopping with legume, contingent crop planning, improved crop varieties, seed and fodder banks, zero tillage/minimum tillage, agro-forestry, concentrate feeding for livestock, fodder management and integrated pest management are some promising climate smart practices [67].

CSA practices and technologies adopted include the following:

1. Improved crop varieties for higher yield.
2. Varieties suitable to cope with drought.
3. Direct-seeded rice.
4. In situ moisture conservation.
5. Water harvesting and its storage.
6. Laser land levelling.
7. Practicing minimum tillage by using zero-till drill or a happy seeder.
8. Nutrient management by using green seeker, leaf colour chart and chlorophyll metre.
9. Managing irrigation by using tensiometer.
10. Accessing weather information through SMS.
11. Residue retention.

12. Site-specific nutrient management.
13. Legume integration and cropping system diversification.
14. Use of solar pump.
15. Use of a mobile phone app helps to calculate how much fertilizer to apply throughout the growing season.

2.5. Climate smart agriculture in India

Climate change and its variability are emerging as the major challenges influencing the performance of agriculture and its sustainability [68]. Long-term changes in shifting weather patterns result in changing climate, which threaten agricultural productivity through high- and low-temperature regimes, increased rainfall variability and rising sea levels that potentially deteriorate coastal freshwater reserves and increase the risk of flooding. The challenge becomes pronounced in the case of unusual or extreme changes in climate, typified by droughts, floods, heat waves, and so on, the frequencies of which are predicted to increase in the future [69]. Developing countries, like India, are more vulnerable to such shocks because of their heavy dependence on agriculture and lack of technical and financial resources to cope up with them [70]. Climate change (and global warming) impacts all sectors of human life. In this context, ICAR started NICRA project to overcome the climate change vulnerability.

2.6. National initiative for climate-resilient agriculture (NICRA)

This programme was started by ICAR during February 2011 with the following objectives:

a. Increase the resilience of Indian agriculture to climate change and climatic variability by development and implementation of production and risk management strategies.
b. Demonstration of site-specific technologies on farmer's field.
c. Enhancing the capacity of scientists and other stakeholders for climate-resilient agriculture research and its application.

Smart agricultural practices in India promoted by ICAR are as follows:

1. Rejuvenation of farming in cyclone- and flood-prone coastal agro-ecosystems through land shaping,
2. Staggered paddy nursery as a contingency measure for drought,
3. Water efficient direct-seeded rice cultivation technology,
4. Drum-seeded rice for improving water use efficiency,
5. Short-duration rice cultivar for drought tolerance,

6. Drought-tolerant short-duration finger millet varieties for late season drought in south interior Karnataka,
7. Short-duration crop varieties suitable for late sowings,
8. Crop diversification for livelihood and food security, sustainability,
9. Flood-tolerant varieties impart resilience to farmers in flood-prone areas,
10. Improving the resilience of poor farmers reclaiming cultivable wastelands,
11. Community tanks/ponds as a means of augmentation and management of village level water resources,
12. Individual farm ponds for improving livelihoods of small farmers,
13. Jalkund—low-cost rainwater-harvesting structures,
14. Check dam-storing excess runoff in streams,
15. Rainwater harvesting and recycling through temporary check dam,
16. Enhancing resilience through improvement in conveyance efficiency,
17. Recharge of wells to improve shallow aquifers,
18. Integrated Farming System modules,
19. Captive rearing of fish seed—a livelihood opportunity in flood-prone areas,
20. Management practices to tackle cold stress in backyard poultry,
21. Shelter management for small ruminants to tackle heat stress and rain storm,
22. Small-farm mechanization through Custom Hiring Centres for farm machinery,
23. Improved planting methods for increasing water use efficiency and crop productivity,
24. Zero-till drill wheat to escape problem of terminal heat stress,
25. In situ incorporation of biomass and crop residues for enhancing soil microbial population and soil health,
26. Village level seed banks for reducing seed shortages,
27. Improved fodder cultivars to solve the problem of fodder scarcity.

3. Conclusion

Climate smart strategies like choice of suitable crop and cultivars, integrated farming system, site-specific nutrient management, residue management, intercropping with legume, conservation agriculture-based resource conservation technology, agro-forestry and crop diversification can help minimize negative impacts to some extent and strengthen farmers by sustainably increasing productivity and income. Location-specific medium-range weather forecasting will play a major role in designing agricultural practices especially in the event of extreme climatic event like high rainfall, drought, frost, hailstorms and heat waves. Water-saving technologies and water-harvesting structures to enhance the availability of water at critical stages of crop growth will be important practice in chronically water-deficient areas. Crop insurance will provide economic security in case of heavy crop loss due to climatic extremes like flood, drought and hailstorms especially to small and marginal farmers. In general, the CSA options integrate traditional and innovative practices, technologies and services that are relevant for particular location. Thus, to meet food security we need such smart agricultural practices which are sustainable, economic and environmentally sound.

4. Future thrust

Precise and accurate weather forecasting for different location will help to make contingent plan for different crop and cropping systems. Researches on precise water, nutrient and pesticide application technologies suitable for small and marginal farmers are needed. Researches on crop residue management, minimum tillage and mulches need to be developed. Development of IFS models is done for different locations keeping in view of farmers resources in that locality. Breeding and biotechnological approaches for crop and varieties need to be developed for adoption in changing climatic scenario.

Acknowledgements

The authors would like to thank the Banaras Hindu University. We also thank the editors for giving opportunities to write the manuscript.

Author details

Mona Nagargade*, Vishal Tyagi and Manoj Kumar Singh

*Address all correspondence to: monanagargade@gmail.com

Department of Agronomy, Institute of Agricultural Sciences, Banaras Hindu University, Varanasi, India

References

[1] IPCC. Summary for policymakers. Climate Change 2014: Impacts, Adaptation, and Vulnerability. Part A: Global and Sectoral Aspects. In: Contribution of Working Group II to the Fifth Assessment Report of the Intergovernmental Panel on Climate Change. Cambridge, United Kingdom and New York, NY, USA: Cambridge University Press; 2014. pp. 1-32

[2] FAO. Climate change, Agriculture and Food Security. Rome, Italy; Food and Agriculture Organization of the United Nations (FAO); 2016

[3] United Nations. Department of Economic and Social Affairs Population Division. World Population Prospects: The 2015 Revision. Working Paper No. ESA/P/WP.241. New York, NY: The Department of Economic and Social Affairs of the UN Secretariat; 2015

[4] Alexandratos N, Bruninsma J. World agriculture towards 2030/2050: The 2012 revision. ESA Working Paper No. 12-03. Rome, Italy: Food and Agriculture Organisation of the United Nations; 2012

[5] Brown ME, Antle JM, Backlund P, Carr ER, Easterling WE, Walsh MK, Ammann C, Attavanich W, Barrett CB, Bellemare MF, Dancheck V, Funk C, Grace K, Ingram JSI, Jiang H, Maletta H, Mata T, Murray A, Ngugi M, Ojima D, O'Neill B, Tebaldi C. Climate Change, Global Food Security, and the U.S. Food System. USDA; 2015. pp. 146

[6] IPCC: Intergovernmental Panel on Climate Change. Six assessment report. 2016

[7] Kumar R, Gautam HR. Climate change and its impact on agricultural productivity in India. Journal of Climatology and Weather Forecasting. 2014;**2**:109. DOI: 10.4172/2332-2594.1000109

[8] Ziska L, Crimmins A, Auclair S, DeGrasse JF, Garofalo AS, Khan I, Loladze AA, Pérez de León A, Showler J, Thurston Walls I. 2016: Ch. 7: Food Safety, Nutrition, and Distribution. The Impacts of Climate Change on Human Health in the United States: A Scientific Assessment. Washington, DC: U.S. Global Change Research Program; 2014. pp. 189-216

[9] Hatfield JG, Takle R, Grotjahn P, Holden RC, Izaurralde T, Mader E, Marshall, Liverman D. In: Melillo JM, Terese Richmond TC, Yohe GW, editors. Ch. 6: Agriculture. Climate Change Impacts in the United States: The Third National Climate Assessment. U.S. Global Change Research Program. 2014. pp. 150-174

[10] Porter JR, Xie L, Challinor AJ, Cochrane K, Howden SM, Iqbal MM, Lobell DB, Travasso MI. Food security and food production systems. In: Field et al. editors. Climate Change 2014: Impacts, Adaptation, and Vulnerability. Part A: Global and Sectoral Aspects. Contribution of Working Group II to the Fifth Assessment Report of the Intergovernmental Panel on Climate Change. Cambridge, UK and New York, USA: Cambridge University Press; 2014. pp. 485-533

[11] CCSP. The Effects of Climate Change on Agriculture, Land Resources, Water Resources, and Biodiversity in the United States. A Report by the U.S. Climate Change Science Program and the Subcommittee on Global Change Research. In: Backlund P, Janetos A, Schimel D, Hatfield J, Boote K, Fay P, Hahn L, Izaurralde C, Kimball BA, Mader T, Morgan J, Ort D, Polley W, Thomson A, Wolfe D, Ryan M, Archer S, Birdsey R, Dahm C, Heath L, Hicke J, Hollinger D, Huxman T, Okin G, Oren R, Randerson J, Schlesinger W, D. Lettenmaier, D. Major, L. Poff, S. Running, L. Hansen, Inouye D, Kelly BP, Meyerson L, Peterson B, Shaw R. Washington, DC, USA: U.S. Environmental Protection Agency; 2008

[12] FAO. Rome: The State of World Fisheries and Aquaculture. World review of fisheries and aquaculture; 2012

[13] Romero-Lankao P, Smith JB, Davidson DJ, Diffenbaugh NS, Kinney PL, Kirshen P, Kovacs P, Villers Ruiz L. North America. In: Barros VR, Field CB, Dokken DJ, Mastrandrea MD, Mach KJ, Bilir TE, Chatterjee M, Ebi KL, Estrada YO, Genova RC, Girma B, Kissel ES, Levy AN, MacCracken S, Mastrandrea PR, White LL, editors. Climate Change 2014: Impacts, Adaptation, and Vulnerability. Part B: Regional Aspects. Contribution of Working Group II to the Fifth Assessment Report of the Intergovernmental Panel on Climate Change. Cambridge, United Kingdom and New York, NY, USA: Cambridge University Press, IPCC; 2014. pp. 1439-1498

[14] FAO. Climate Change Guidelines for Forest Managers. FAO Forestry Paper 172. Rome; 2013

[15] Singh G. Agriculture diversification for food, nutrition, livelihood and environmental security: Challenges and opportunities. Indian Journal of Agronomy. 2015;**60**(2):172-184

[16] Mall RK, Gupta A, Singh R, Singh RS, Rathore LS. Water resources and climate change: An Indian perspective. Current Science. 2005;**90**:1610-1626

[17] Mall RK, Singh R, Gupta A, Singh RS, Srinivasan G. Impact of climate change on Indian agriculture: A review. Climate Change. 2006;**78**:445-478

[18] CCAFS. Big Facts on Climate Change, Agriculture and Food Security. Copenhagen, Denmark: CGIAR Research Program on Climate Change, Agriculture and Food Security (CCAFS); 2013

[19] FAO. Climate Smart Agriculture: Policies, Practices and Financing For Food Security, Adaptation and Mitigation. Rome, Italy: Food and Agriculture Organization of the United Nations (FAO); 2010

[20] Sapkota TB, Jat ML, Aryal JP, Jat RK, Khatri-Chhetri A. Climate change adaptation, greenhouse gas mitigation and economic profitability of conservation agriculture: Some examples from cereal systems of Indo-Gangetic Plains. Journal of Integrative Agriculture. 2015;**14**(8):1524-1533

[21] Mittal S. Modern ICT for Agricultural Development and Risk Management in Smallholder Agriculture in India. CIMMYT. Socio-Economics Working Paper 3. Mexico, D.F.: CIMMYT; 2012

[22] Kumar V, Ladha JK. Direct seeding of rice: Recent developments and future research needs. Advances in Agronomy. 2011;**111**:297-413

[23] Jugsujinda A, Delaune RD, Lindau CW. Factors controlling carbon dioxide and methane production in acid sulfate soils. Water Air and Soil Pollution. 1996;**87**:345-355 EPA. Carbon Sequestration in Agriculture and Forestry. 2008. Available from: www.epa.gov/sequestration/index.html

[24] Philippot L, Hallin S. Towards food, feed and energy crops mitigating climate change. Trends in Plant Science. 2011;**16**:476-480

[25] Barfoot P, Brookes G. Key global environmental impacts of genetically modified (GM) crop use 1996-2012. GM Crops and Food: Biotechnology in Agriculture and the Food Chain. 2014;**5**(2):149-160

[26] Lybbert T, Sumner D. Agricultural technologies for climate change mitigation and adaptation in developing countries: Policy options for innovation and technology diffusion. ICTSD-IPC Platform on Climate Change, ATS Policy Brief 6. 2010. Available from: http://ictsd.org/i/publications/77118/

[27] Shrawat AK, Good AG. Genetic engineering approaches to improving nitrogen use efficiency. ISB Report. May 2008. Available from: http://www.isb.vt.edu/news/2008/news08.may.htm#may0801. 2008

[28] Subbarao GV, Ban T, Kishii M, Ito O, Samejima HY, Wang SJ, Pearse S, Gopalakrishnan K, Nakahara AKM, Zakir Hossain H, Tsujimoto WL, Berry W. Can biological nitrification inhibition (BNI) genes from perennial *Leymus racemosus* (Triticeae) combat nitrification in wheat farming? Plant Soil. 2007;**299**:55-64

[29] Lutz D. Creating plants that make their own fertilizer. Washington University in St. Louis News Release. 22 August 2013. Available from: http://news.wustl.edu/news/Pages/25585.aspx.

[30] OECD. Synthetic biology: A new and promising technology. In: OECD, Emerging Policy Issues in Synthetic Biology. Paris: OECD Publishing; 2014. DOI: http://dx.doi.org/10.1787/9789264208421-4-en

[31] Alford AR, Hegarty RS, Parnell PF, Cacho OJ, Herd RM, Griffith GR. The impact of breeding to reduce residual feed intake on enteric methane emissions from the Australian beef industry. Australian Journal of Experimental Agriculture. 2006;**46**:813-820

[32] Dwivedi SL, Sahrawat K, Upadhyaya H, Ortiz R. Food, nutrition and agrobiodiversity under global climate change. Advances in Agronomy. 2013;**120**:1-118

[33] Hasanuzzaman M, Nahar K, Alam MdM, Roychowdhury R, Fujita M. Physiological, biochemical, and molecular mechanisms of heat stress tolerance in plants. International Journal of Molecular Sciences. 2013;**14**:9643-9684

[34] Tuteja N. Integrated calcium signaling in plants. In: Baluska F, Mancuso S, editors. Signaling in Plants I. Heidelberg, Germany: Springer; 2009. pp. 29-49

[35] Kaur N, Gupta AK. Signal transduction pathways under abiotic stresses in plants. Current Science. 2005;**88**:1771-1780

[36] Gusta L. Abiotic stresses and agricultural sustainability. Journal of Crop Improvement. 2012;**26**:415-427

[37] Katiyar-Agarwal S, Agarwal M, Grover A. Heat-tolerant basmati rice engineered by over-expression of hsp101. Plant Molecular Biology. 2003;**51**:677-686

[38] Kuromori T, Mizoi J, Umezawa T, Yamaguchi-Shinozaki K, Shinozaki K. Drought stress signaling network. Molecular Biology. 2014;**2**:383-409

[39] Todaka D, Shinozaki K, Yamaguchi-Shinozaki K. Recent advances in the dissection of drought-stress regulatory networks and strategies for development of drought-tolerant transgenic rice plants. Frontiers in Plant Science. 2015;**6**:1-20

[40] Golldack D, Li C, Mohan H, Probst N. Tolerance to drought and salt stress in plants: Unraveling the signaling networks. Frontiers in Plant Science. 2014;**5**:151

[41] Bhardwaj J, Yadav SK. Genetic mechanisms of drought stress tolerance, implications of transgenic crops for agriculture. In: Lichtfouse E, editor. Agroecology and Strategies for Climate Change: Springer; 2012. pp. 213-235

[42] Karaba A, Dixit S, Greco R, Aharoni A, Trijatmiko KR, Marsch-Martinez N, Krishnan A, Nataraja KN, Udayakumar M, Pereira A. Improvement of water use efficiency in rice by expression of HARDY, an Arabidopsis drought and salt tolerance gene. Proceedings of the National Academy of Sciences of the United States of America. 2007;**104**:15270-15275

[43] Castiglionit P, Warner D, Bensen RJ, Anstrom DC, Harrison J, Stoecker M, Abad M, Kumar G, Salvador S, D' Ordine R, Navarro S, Back S, Fernandes M, Targolli J, Dasgupta S, Bonin C, Luethy MH, Heard JE. Bacterial RNA chaperones confer abiotic stress tolerance in plants and improved grain yield in maize under water–limited conditions. Plant Physiology. 2008;**147**:446-455

[44] FAO. Aquastat. 2009. Available from: www.fao.org/nr/water/aquastat/data/query/index.html

[45] Tester M, Davenport R. Na^+ tolerance and Na^+ transport in higher plants. Annals of Botany. 2003;**91**:503-527

[46] Chinnusaamy V, Jagendorf A, Zhu JK. Understanding and improving salt tolerance in plants. Crop Science. 2005;**45**:437-448

[47] Ortiz R, Iwanaga M, Reynolds MP, Wu X, Crouch JH. Overview on crop genetic engineering for drought-prone environments. Journal of Semi-Arid Tropical Agricultural Research. 2007;**4** pp. 1-30. . Available from: http://www.icrisat.org/journal/SpecialProject/sp3.pdf

[48] He C, Yan J, Shen G, Fu L, Holaday AS, Auld D, Blumwald E, Zhang H. Expression of an Arabidopsis vacuolar sodium/proton antiporter gene in cotton improves photosynthetic performance under salt conditions and increases fiber yield in the field. Plant Cell Physiology. 2005;**46**:1848-1854

[49] Moghaieb RE, Nakamura A, Saneoka H, Fujita K. Evaluation of salt tolerance in ectoine-transgenic tomato plants (*Lycopersicon esculentum*) in terms of photosynthesis, osmotic adjustment, and carbon partitioning. GM Crops. 2011;**2**:58-65

[50] Ortiz R, Jarvis A, Fox P, Aggarwal PK, Campbell BM. Plant genetic engineering, climate change and food security. CCAFS Working Paper No. 72. CGIAR Research Program on

Climate Change, Agriculture and Food Security (CCAFS). Copenhagen, Denmark. 2014. Available online at: www.ccafs.cgiar.org

[51] Dar MH, Singh S, Zaidi NW, Shukla S. Sahbhagi Dhan: Science's answer to drought problems. STRASA News. 2012;**5**:1-13

[52] Singh A, Yadav A, Matthew K, Varrieur V, Singh S. Success of IRRI stress tolerant rice varieties in Bihar, India-Experience of IRRAS farmers. In: Poster Presented at Fourth International Rice Conference; 27 October to 01 November 2014; Bangkok, Thailand. 2014a

[53] Haefelea SM, Kato Y, Singh S. Climate ready rice: Augmenting drought tolerance with best management practices. Field Crops Research. 2016;**190**:60-69

[54] Prasad YG, Maheswari M, Dixit S, Srinivasarao CH, Sikka AK, Venkateswarlu B, Sudhakar N, Prabhu Kumar S, Singh AK, Gogoi AK, Singh AK, Singh YV, Mishra A. Smart Practices and Technologies for Climate Resilient Agriculture. Hyderabad: Central Research Institute for Dryland Agriculture (ICAR); 2014. p. 76

[55] Prasad YG, Maheswari M, Dixit S, Srinivasarao CH, Sikka AK, Venkateswarlu B, Sudhakar N, Prabhu Kumar S, Singh AK, Gogoi AK, Singh AK, Singh YV, Mishra A. Smart Practices and Technologies for Climate Resilient Agriculture. Hyderabad: Central Research Institute for Dryland Agriculture (ICAR); 2014. p. 76

[56] IYP. International Year of Pulses, Theme: Productivity & Environmental Sustainability. 2016. Available from: http://pulses.org/future-of-food/pulses-sustainable-food

[57] Singh SS, Kumar S, Haris AA, Singh AK, Sangle UR, Bhatt BP, Yadav A, Singh US, Singh S. Performance of lentil succeeding rice in puddle and un-puddled transplanted conditions in drought-prone rainfed ecosystem. In: Poster presented in Fourth International Rice Conference; 27 October to 1 November 2014; Bangkok, Thailand. 2014

[58] Jat M, Gathala MK, Ladha JK, Saharawat YS, Jat AS, Kumar Vipin, Sharma SK, Kumar V, Gupta R. Evaluation of precision land leveling and double zero-till systems in the rice–wheat rotation: Water use, productivity, profitability and soil physical properties. Soil & Tillage Research. 2009;**105**:112-121

[59] Bhattacharyya R, Das TK, Sudhishri S, Dudwal B, Sharma AR, Bhatia A, Singh G. Conservation agriculture effects on soil organic carbon accumulation and crop productivity under a rice–wheat cropping system in the western Indo-Gangetic Plains. European Journal of Agronomy. 2015;**70**:11-21

[60] Jat RK, Sapkota TB, Singh RG, Jat MLL, Kumar M, Gupta RK. Seven years of conservation agriculture in a rice-wheat rotation of Eastern Gangetic Plains of South Asia: Yield trends and economic profitability. Field Crops Research. 2014;**164**:199-210

[61] Grace PR, Jain MC, Harrington GP, Antle J, Aggarwal PK, Ogle S, Pautian K, Basso B. Longterm sustainability of the tropical and subtropical rice-wheat system: An environmental perspective. In: Ladha JK, Hill J, Gupta RK, Duxbury J, Buresh RJ, editors.

Improving the Productivity and sustainability of Rice-Wheat systems: Issues and Impact. ASA. 2003. Special Publications **65**:1-18

[62] Gupta DK, Bhatia A, Kumar A, Das TK, Jain N, Tomer R, Malyan SK, Fagodiya RK, Dubey R, Pathak H. Mitigation of greenhouse gas emission from rice–wheat system of the Indo-Gangetic plains: Through tillage, irrigation and fertilizer management. Agriculture, Ecosystems and Environment. 2016;**230**:1-9

[63] Mahajan A, Sharma R. Integrated nutrient management (INM) system-Concept, need and future strategy. Agrobios Newsletter. 2005;**4** (3):29-32

[64] Prasad JVNS, Rao CS, Srinivas K, Jyothi CN, Venkateswarlu B, Ramachandrappa BK, Dhanapal GN, Ravichandra K, Mishra PK. Effect of ten years of reduced tillage and recycling of organic matter on crop yields, soil organic carbon and its fractions in Alfisols of semi arid tropics of southern India. Soil & Tillage Research. 2016;**156**:131-139

[65] Bhatia A, Pathak H, Jain N, Singh PK, Tomer R. Greenhouse gas mitigation in rice–wheat system with leaf color chart-based urea application. Environmental Monitoring and Assessment. 2012;**184**:3095-3107

[66] Rathore LS, Chattopadhyay N, Chandras SV. Combating effect of climate change on Indian Agriculture through smart weather forecasting and ICT application. Journal of Climate Change. 2016;**2**(1):43-51

[67] Khatri-Chhetri A, Aggarwal PK, Joshi PK, Vyas S. Farmer's prioritization of climate smart agriculture (CSA) technologies agricultural system. Agricultural Systems 02197. 2016:8. DOI: http://dx.doi.org/10.1016/j.agsy.2016.10.005

[68] De Salvo, Raffael R, Moser R. The impact of climate change on permanent crops in an alpine region: A Ricardian analysis. Agricultural System. 2013;**118**:23-32

[69] World Bank. Turn Down the Heat: Climate Extremes, Regional Impacts, and the Case for Resilience. A Report for the World Bank by the Potsdam Institute for Climate Impact Research and Climate Analytics. Washington, DC: World Bank; 2013

[70] Lal PN, Mitchell T, Aldunce P, Auld H, Mechler R, Miyan A, Romano LE, Zakaria S. National systems for managing the risks from climate extremes and disasters. In: Field CB, Barros V, Stocker TF, Qin D, Dokken DJ, Ebi KL, Mastrandrea MD, Mach KJ, Plattner GK, Allen SK, Tignor M, Midgley PM, editors. Managing the Risks of Extreme Events and Disasters to Advance Climate Change Adaptation. A Special Report of Working Groups I and II of the Intergovernmental Panel on Climate Change. Cambridge and New York: Cambridge University Press; 2012. pp. 339-392

Chapter 9

Adaptive Management Framework for Evaluating and Adjusting Microclimate Parameters in Tropical Greenhouse Crop Production Systems

Redmond R. Shamshiri, Muhammad Razif Mahadi,
Kelly R. Thorp, Wan Ishak Wan Ismail,
Desa Ahmad and Hasfalina Che Man

Additional information is available at the end of the chapter

http://dx.doi.org/10.5772/intechopen.69972

Abstract

High operational costs of greenhouse production in hot and humid climate condition due to the initial investments on structure, equipment, and energy necessitate practicing advanced techniques for more efficient use of available resources. This chapter describes design and concepts of an adaptive management framework for evaluating and adjusting optimality degrees and comfort ratios of microclimate parameters, as well as predicting the expected yield in greenhouse cultivation of tomato. A systematic approach is presented for automatic data collection and processing with the objective to produce knowledge-based information in achieving optimum microclimate for high-quality and high-yield tomato. Applications of relevant computer models are demonstrated through case-study examples for use in an iterative way to simulate and compare different scenarios. The presented framework can contribute to future studies for providing best management decisions such as site selection, optimum growing season, scheduling efficiencies, energy management with different climate control systems, and risk assessments associated with each task.

Keywords: greenhouse, climate control, microclimate evaluation, tomato, ventilation, evaporative cooling

1. Introduction

The increasing market demand for high-quality food products have replaced open-field cultivations of Solanaceae and Cucurbits crops with modern plant production systems for more efficient use of available resources. Closed-field cultivations by means of commercial greenhouses have been changed over the last three decades from basic structures to advanced controlled environments for optimizing plant's productivity and producing high yields at low expenses. The higher costs of greenhouse operation due to the initial investments on structure, equipment, and energy necessitate practicing advanced techniques of automation for efficient control of the microenvironment. Research trends in this field are directed toward developing innovative solutions for shifting from energy-consuming to energy neutral greenhouses with the ultimate objective of increasing profits. This is, however, challenging due to the lack of accurate information about interactions between crops and environment at different growth stages, as well as the complexity of the dynamic system that is subjected to changes with internal and external factors. Plant-based engineering has helped researchers with proper management policies to embrace these uncertainties through modeling and integrated-learning approaches. Several uncertainties with greenhouse cultivation include climate variability, expected yield, optimum references of microclimate parameters, comfort ratios, insecurity of resources, complexity of the system states, lack of accurate information about interactions between plants and environment, and the relationships between biological and ecological system.

Greenhouse microclimate control has been a large field of study for many years. Much work has been done for moderate cold climate conditions as opposed to tropical lowlands. In contrast to cold arid climate, the main objective of a greenhouse in hot and humid regions such as lowlands of south-east Asia (**Figure 1**) is to protect crop against fluctuations of external conditions such as extreme winds, heavy seasonal rainfalls, typhoons, extreme solar radiation, occasional water shortage, high air temperature, high humidity, and invasion of pests and diseases. The major concern with greenhouses in these regions is the crop stress due to the adverse microclimate that reduces plant evapotranspiration rate and causes production failure. Evaporative cooling systems by means of misting, pad-and-fan, and swamp cooling are widely used in tropical greenhouses of south-east Asia for manipulating crop growth microclimate; however, these systems have not reached their optimum potential due to their conventional automation and control methods. If properly managed, tropical greenhouses can provide suitable growth condition for tomato cultivation by maintaining inside microclimate close to the outside, with an expected yield that varies between 30 and over 100 tons/ha (vs. open-field yield of 15–30 tons/ha) depending on soil culture or hydroponics medium.

Profitability and investment returns of commercial greenhouses are tightly linked to management decisions. One of the main factors to be considered in this context is the sustainability of

Figure 1. Outside and inside view of tropical greenhouses in the lowlands of Malaysia.

operations through proper management of available resources. Modern greenhouses are required to exhibit integration of automation, cultural practices, and environmental control using object-oriented analysis of the subsystems. The primary concepts and methods of automation-culture-environment system analysis (ACESYS) in controlled environments plant production (CEPP) have been introduced and expanded in the works of [1–3]. Some of the earliest examples of object-oriented analysis and modeling applications including optimal control strategies and decision-support software in advanced CEPP systems can be found in the works of [2, 4–6]. The purpose of object-oriented system analysis approach according to Ref. [3] is to develop a set of foundation classes that can be used to effectively describe the components of the automation system. This, however, requires a comprehensive understanding of the interaction between crop's growth response and environment characteristics. Some of the specific applications and benefits of system analysis in greenhouse production includes integrated energy-efficient strategies, extracting unique and new knowledge that provides valuable insight to local growers and beyond, understanding limitations of resources and balancing between input and output expectancies, improving technology and increasing returns, providing business attraction for local stakeholders, minimizing energy requirements and eliminating tedious operations, increasing production quality and quantity to satisfy market demands, and technology adaptation by balancing between fixed and flexible automation for various crop production. With this perspective, the convolution of several possible scenarios and combination of culture classes (i.e., climate control parameters) and objects (i.e., tomato crop at different growth stages) necessitates computer-based analysis program within the concepts of a systematic framework approach such as adaptive management.

Adaptive management was initially introduced at the University of Florida [7] as an iterative method for managing natural resource in the systems with wide range of responses to management choices and to help manager's difficulty in understanding the systems' dynamics [8] and plant's responses [9]. It is defined as "a systematic process for continually improving management policies and practices by learning from the outcomes of operational programs" [10]. Adaptive management was created based on the needs of environment and ecosystem managers with an iterative processing tool that acknowledged complexity and uncertainty, with a focus on learning and for continuous inputs [11]. It has been widely used as a new design technique for large database that manages and assists the immense data collection, data analysis, and data storage of distributed sets of experiments associated with environmental, meteorological, biological, and medical research problems or other technical and experimental assessments that utilize large-scale data sources within multiple and separate engineering or laboratory facilities. Examples include the work of Refs. [12–15]. The principles of adaptive management according to Ref. [16] suggest using the best available knowledge to design and implement management plans, while establishing an institutional structure that enables learning from outcomes to adjust and improve future decision making. This structured approach is an efficient method in developing decision-support tools for systems design, management, and operation by recognizing the importance of natural variability in contributing to ecological resilience and productivity.

This chapter provides a systematic process of incorporating new and existing knowledge that can be used in developing management decisions for achieving optimum microclimate.

It describes design and concepts of an adaptive management framework for evaluating optimality degrees and comfort ratios of air temperature (T), relative humidity (RH), and vapor pressure deficit (VPD), as well as prediction of the expected yield in greenhouse cultivation of tomato. The presented framework was designed to allow production managers to ask "what-if" type of questions for further quantitative inclusion to avoid possible detriment decisions. It also provides an in-depth rigorous analysis tool for decision making or decision procrastination when facing uncertainties. It can assist in enhancing scheduling efficiency and guiding investments through different simulated scenarios that are based on information analysis to support optimal restoration strategies. In the rest of this chapter, we refer to the term "microclimate parameters T, RH, or VPD" by $\mathcal{M}$. We also use $Opt(\mathcal{M}) = \alpha$, and $Cft(\mathcal{M}, t, \alpha) = \beta$, to refer to the terms "optimality degree" and "comfort ratio," respectively, defined in Sections 3.1 and 3.2.

2. Adaptive greenhouse model

The key to an adaptive greenhouse is the computer model that drives specific implementations of other components. An adaptive management framework for microclimate evaluation and control in greenhouse production systems is proposed in **Figure 2**. A diagram of the steps in the analysis process is shown by rectangles. The arrows are the direction of the process, and the central spiral highlights the goal of arriving at a compromising decision based on a shared set of objectives developed through the iterative process. The three essential elements in this structure are (i) data entry and retrieval, (ii) computer model (expressed by mathematical equations), and (iii) data

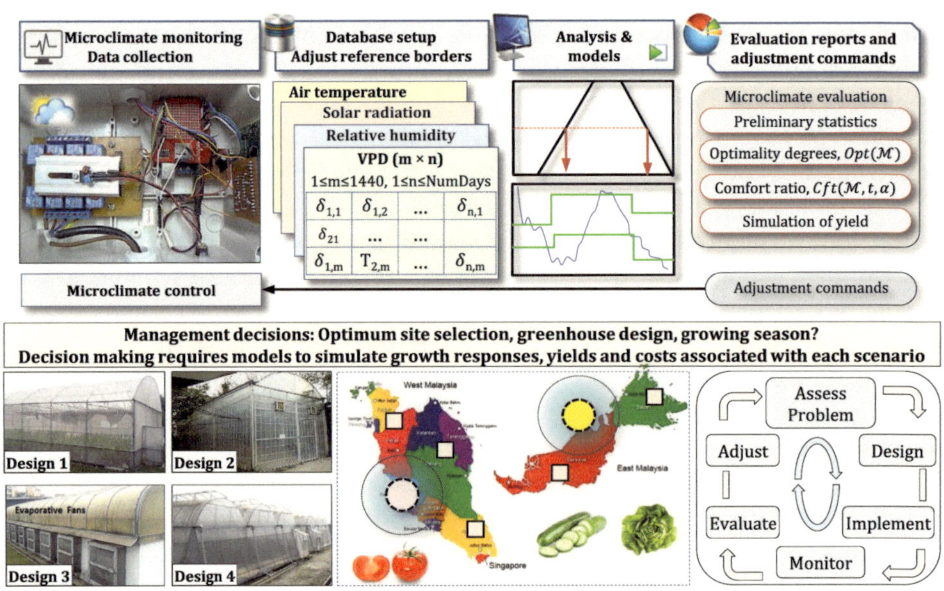

Figure 2. Diagram of the adaptive management framework for monitoring, data processing, evaluating, and adjusting greenhouse microclimate with an iterative analysis approach for scenario analysis with greenhouse crop production.

http://dx.doi.org/10.5772/intechopen.69972

analysis components. The data entry component may be implemented by direct interfacing with real-time data acquisition system or by using web-based and desktop application software. Computer model is application specific; it can be updated and is usually condensed and produced from previous extensive research works in crop physiology. Data analysis comprises implementing relevant techniques within the retrieval component (i.e., programmable spread sheets) or by integrating with third-party applications (i.e., Simulink blocks). The proposed framework can be adapted to new research projects for working with different culture classes and objects by which many specific scenarios may be modeled and analyzed. It carefully monitors the possible outcomes of the system for better understanding of the process in order to adjust control parameters through an iterative learning process.

The framework utilizes a custom-designed data acquisition, and control system [17] that was built using Arduino Uno prototype microcontroller board for monitoring and manipulating of the microclimate parameters. Three computer models were employed by the framework for evaluation and adjusting of optimality degrees $Opt(\mathcal{M})$, comfort ratio $Cft(\mathcal{M}, t, \alpha)$, and prediction of the expected yield. The framework was implemented in MATLAB® (The MathWorks Inc, Natick, MA, USA) environment through Simulink blocks and coding of various main functions and sub-functions that were stored as "m-files." Different toolboxes were developed for the immense data-analyzing tasks as shown in **Figure 3**. The framework structure was designed in a way that end users can create (or update) entries in database, select report type (1-day or multidays report), and proceed with a specific analysis procedure. The database is a dynamic flat file type that can be created by entering collected data, either manually from previously stored sources such as excel sheets or directly from the hardware interface. The computer models presented in this chapter are focused on tomato (*Lycopersicon*

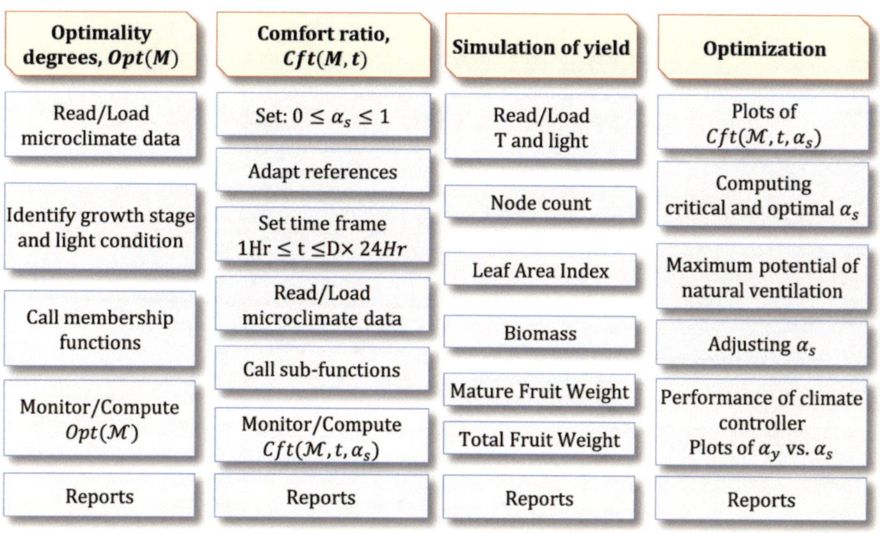

Figure 3. Arrangement of the process in the framework toolboxes.

esculentum); however, with slight modification, the framework can be reprogrammed to work with other greenhouse crops provided that their yield prediction and growth response models are available. Results of microclimate evaluation and set-point manipulation discussed in Sections 3 and 4 can contribute to dynamic greenhouse climate control strategies [18] such as the one in Ref. [19]. An example is provided by comparing a model reference-adaptive greenhouse microclimate controller with conventional closed-loop feedback shown in **Figure 4**. In this scheme, the control law is adapted with the new greenhouse states based on the optimized set points as shown in the diagram of **Figure 5** [19] for a specific microclimate

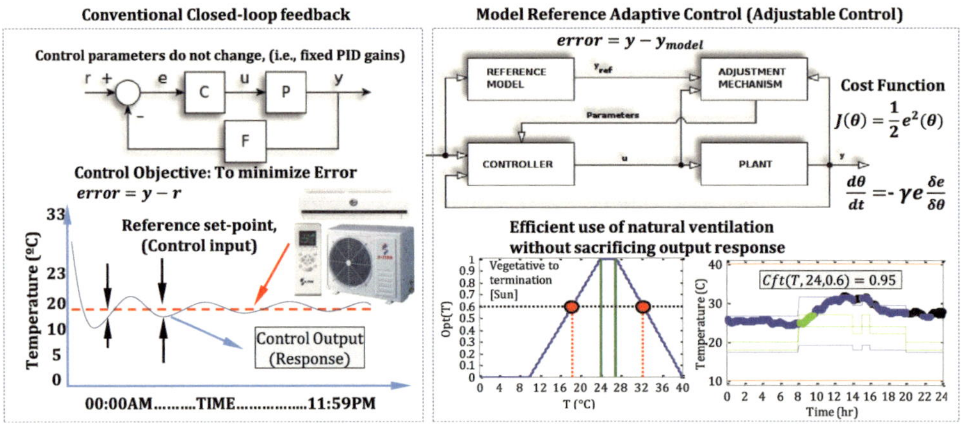

Figure 4. Demonstration of conventional greenhouse controller (left) versus model reference adaptive controller (right).

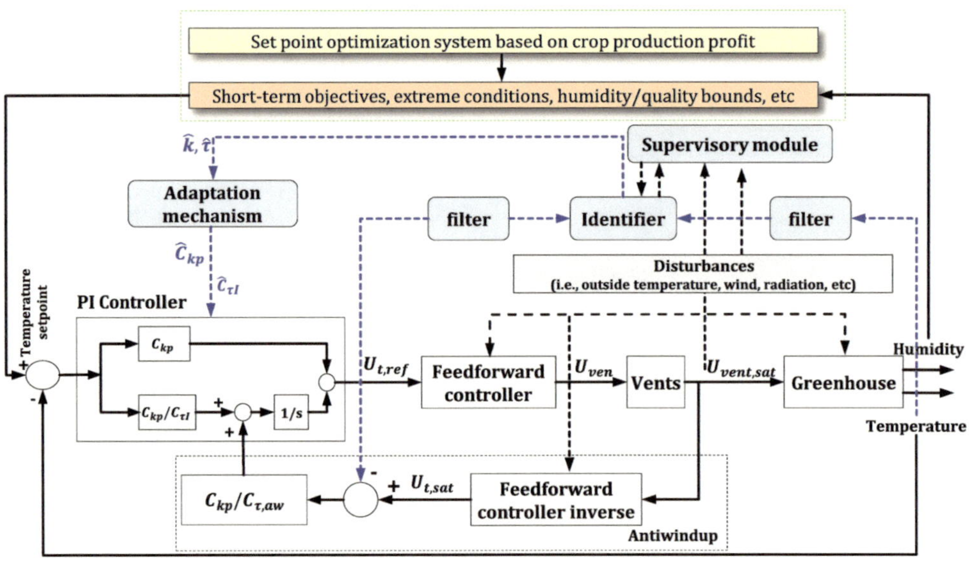

Figure 5. Adaptive control of greenhouse air temperature based on manipulated set point as discussed in Ref. [19].

http://dx.doi.org/10.5772/intechopen.69972

parameter (i.e., air temperature), causing other microclimate parameter (i.e., humidity) to be actually controlled via set-point manipulation.

3. Microclimate evaluation with manipulated set points

3.1. Optimality degrees of microclimate

Optimality degree of a microclimate parameter denoted by $Opt(\mathcal{M}) = \alpha$ is a quantitative value between 0 and 1 that represents how close a microclimate measurement (T, RH, or VPD) is to its ideal value as required by the greenhouse crop at specific growth stage and climate condition. This value can be computed from experimental models that correlate different levels of microclimate parameters with yield and quality of the greenhouse crop. An example of such models is the one developed for air temperature and relative humidity by the Ohio Agricultural Research and Development Center [20, 21]. These models define optimality degrees of T, RH for greenhouse cultivation of tomato with independent trapezoid membership-function growth response plots that are specific for different growth stages and three light conditions (night, sun, and cloud). These plots were originated using utility theory with the goal of simultaneously achieving high-yield and high-quality fruit. The knowledge behind these plots was condensed from extensive scientific literature and peer-reviewed published research on greenhouse tomato production and physiology. Mathematical expressions and plots of membership functions for defining optimality degrees of T and RH are available in Ref. [22]. The sets of membership functions for defining optimality degrees of VPD are presented in the work of Ref. [23]. According to this model, a membership function for specific growth stage and light condition on the universe of discourse is defined as $Opt(\mathcal{M})_{GS,\ (Light)} : \mathcal{M} \rightarrow [0,1]$, where $\mathcal{M} :\ T, RH,\ and\ VPD$ is the universe of discourse (input). In other words, each $\mathcal{M}$ reading in the greenhouse at time $t_{m,n}$, is mapped to a value between 0 and 1 that quantifies its optimality for tomato production. The two indexes m and n refer to specific minute and date of a measurement. In this model, an optimality degree equal to 1 refers to a potential yield with marketable value high-quality fruit. For example, $Opt(T) = 1$ is associated with $T \in [24, 27]°C$ at the vegetative to mature fruiting growth stage during sun hours. For the same growth stage and light condition, a wider reference border, that is, $T \in [18.4, 32.2]°C$, is associated with a lower range of optimality degrees, $Opt(T) \in [0.6, 1]$. In other words, a greenhouse air temperature equal to 32.2°C during sun hours is 60% optimal for tomato production in the vegetative to mature fruiting growth stage. The reference values corresponding to the optimal, marginal, and failure T and RH are summarized in **Table 1**. These values for VPD depend on the range of T and RH and are discussed in Ref. [23]. The optimality-degree model was implemented in the framework as a toolbox and was successfully used in evaluating microclimate parameters. Results of an actual case study on a net-screen-covered greenhouse in tropical lowlands of Malaysia are provided in **Figures 6** and **7** [22].

3.2. Comfort ratio of microclimate

Comfort ratio of a microclimate parameter, denoted by $Cft(\mathcal{M}, t, \alpha_s)_{GS} = \beta$, is defined as the percentage of $\mathcal{M}$ data collected during time frame t that falls inside reference borders of $\mathcal{M}$

Temperature			Relative humidity		
Growth stage	**Reference border**	**Value (°C)**	**Growth stage**	**Reference border**	**Value (%)**
Early growth (GS1)	$T1_{\alpha_0 L}$	9	**Early growth (GS1)**	$RH1_{\alpha_0 L}$	60
	$T1_{\alpha_0 H}$	35		$RH1_{\alpha_1 L}$	75
	$T1_{\alpha_1 L}$	24		$RH1_{\alpha_1 H}$	99
	$T1_{\alpha_1 H}$	26.1			
Vegetative to termination (GS2-5)	$T2_{\alpha_0 L}$	10	**Vegetative (GS2)**	$RH2_{\alpha_0 L}$	40
	$T2_{\alpha_0 H}$	40		$RH2_{\alpha_0 H}$	99
	$T2_{\alpha_{0.5} \mathrm{N}}$	17		$RH2_{\alpha_1 L}$	70
	$T2_{\alpha_1 \mathrm{L,N}}$	18		$RH2_{\alpha_1 H}$	80
	$T2_{\alpha_1 H,N}$	20	**Flowering to termination (GS3-5)**		
	$T2_{\alpha_1 \mathrm{L,S}}$	24		$RH3_{\alpha_0 L}$	30
	$T2_{\alpha_1 H,S}$	27		$RH3_{\alpha_0 H}$	99
	$T2_{\alpha_1 L,C}$	22		$RH3_{\alpha_1 L}$	60
	$T2_{\alpha_1 H,C}$	24		$RH3_{\alpha_1 H}$	80

Indices are: L: lower border, H: higher border, N: night, C: cloud, S: sun, α_0: index of failure, $\alpha_{0.5}$: index of Opt=0.5, α_1: index of Opt=1.

Table 1. Reference values of optimal and failure T and RH at different growth stages and light conditions.

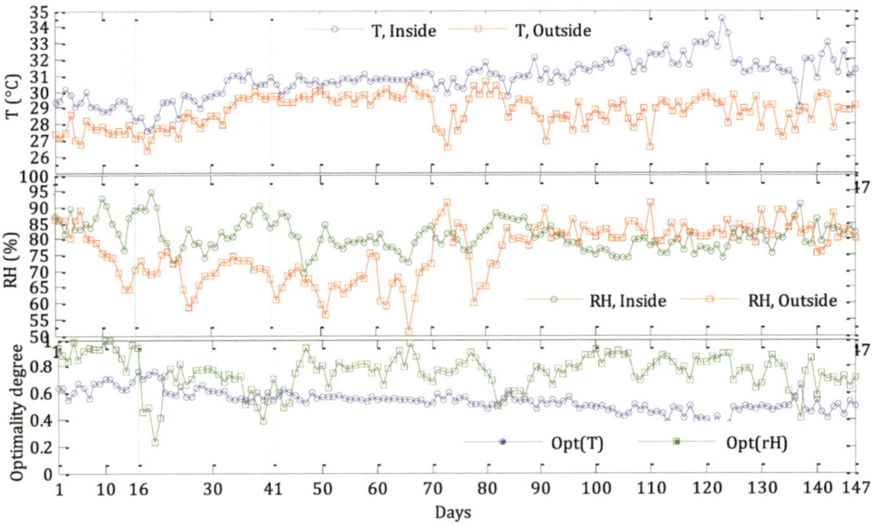

Figure 6. Plots of daily averaged air temperature, RH, and associated optimality degrees from a tropical greenhouse experiment (Source: [22]).

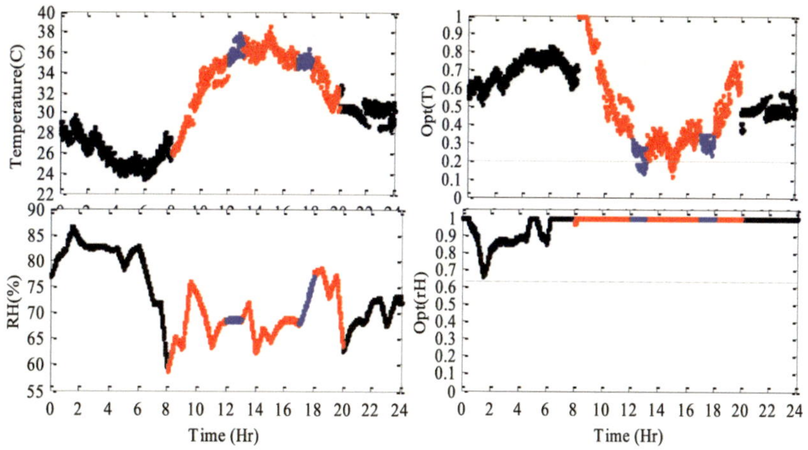

Figure 7. Demonstration of real-time measured air temperature and RH (left) and corresponding optimality degrees (right) for a random cultivation day at the flowering to mature fruiting growth stage (date: March 11, 2015) in a tropical greenhouse. Each color represents a light condition, back: night, red: sun, blue: cloud (Source: [22]).

associated with α_s at a specific growth stage. A 100% ideal microclimate growth condition is therefore defined as $Cft(\mathcal{M}, t, 1) = 1$. The notation α_s refers to user-preferred optimality degree for adjusting the reference borders that is desired for microclimate evaluation or control. The reference borders for a given α_s are calculated from available simulation models (i.e., from the membership function growth response models of [21, 23]). For the purpose of this chapter, mathematical descriptions of Ref. [21] model for defining reference borders of air temperature and relative humidity are adapted and provided in **Table 2**. An example is demonstrated in **Figure 8** for constructing reference borders of air temperature associated with $\alpha_s = 0.8$ at the vegetative to mature fruiting growth stage. The procedure is similar for other microclimate parameters (RH and VPD) at other growth stages and for any selection of $0 \leq \alpha_s \leq 1$. The framework algorithm automatically selects proper membership functions from database according to the light condition and growth stage and computes the reference borders for the given α_s. The light condition in this demonstration belongs to a random day, date: December 15, 2013. The reference borders corresponding to $\alpha_s = 0$, $\alpha_s = 0.8$ and $\alpha_s = 1$ are shown in red, blue, and green colors, respectively. The framework plots data inside each reference border in different colors (black for $\alpha_s = 0$, blue for a preferred α_s, and green for $\alpha_s = 1$). If a measurement lies outside marginal reference borders ($\alpha_s = 0$), it will be plotted in red.

The main purpose of introducing comfort ratio and corresponding graphical demonstration is to address deviation of microclimate responses with respect to different reference borders and to compare it for different cultivation days or greenhouse designs. A practical example is provided in **Figure 9** for air temperature collected from a naturally ventilated greenhouse in two random days, one at the early growth and the other at the mature fruiting stage. The reference borders associated with a preferred optimality degree (i.e., $\alpha_s = 0.7$) are shown in blue color-dashed lines. Moreover, the reference borders corresponding to failure air

Reference function		Preferred optimality
	$T1_{\alpha_0 L} \wedge T1_{\alpha_0 H}$	$\alpha = 0$
$T(\alpha)_{G1A} =$	$\alpha(T1_{\alpha_1 L} - T1_{\alpha_0 L}) + T1_{\alpha_0 L} \wedge \alpha(T1_{\alpha_1 H} - T1_{\alpha_0 H}) + T1_{\alpha_0 H}$	$0 < \alpha < 1$
	$[T1_{\alpha_1 L}, T1_{\alpha_1 H}]$	$\alpha = 1$
	$T2_{\alpha_0 L} \wedge T2_{\alpha_0 H}$	$\alpha = 0$
$T(\alpha)_{G2S} =$	$\alpha(T2_{\alpha_1 L,S} - T2_{\alpha_0 L}) + T2_{\alpha_0 L} \wedge \alpha(T2_{\alpha_1 H,S} - T2_{\alpha_0 H}) + T2_{\alpha_0 H}$	$0 < \alpha < 1$
	$[T2_{\alpha_1 L,S}, T2_{\alpha_1 H,S}]$	$\alpha = 1$
	$T2_{\alpha_0 L} \wedge T2_{\alpha_0 H}$	$\alpha = 0$
$T(\alpha)_{G2C} =$	$\alpha(T2_{\alpha_1 L,C} - T2_{\alpha_0 L}) + T2_{\alpha_0 L} \wedge \alpha(T2_{\alpha_1 H,C} - T2_{\alpha_0 H}) + T2_{\alpha_0 H}$	$0 < \alpha < 1$
	$[T2_{\alpha_1 L,C}, T2_{\alpha_1 H,C}]$	$\alpha = 1$
	$T2_{\alpha_0 L} \wedge T2_{\alpha_0 H}$	$\alpha = 0$
	$2\alpha(T2_{\alpha_{0.5} N} - T2_{\alpha_0 L}) + T2_{\alpha_0 L}$	$0 < \alpha < 0.5$
$T(\alpha)_{G2N} =$	$T2_{\alpha_{0.5} N}$	$\alpha = 0.5$
	$2\alpha(T2_{\alpha_1 L,N} - T2_{\alpha_{0.5} N}) + T2_{\alpha_{0.5} N} - (T2_{\alpha_1 L,N} - T2_{\alpha_{0.5} N})$	$0.5 < \alpha < 1$
	$[T2_{\alpha_1 L,N}, T2_{\alpha_1 H,N}]$	$\alpha = 1$
	$\alpha(T2_{\alpha_1 H,N} - T2_{\alpha_0 H}) + T2_{\alpha_0 H}$	$0 < \alpha < 1$
	$RH1_{\alpha_0 L}$	$\alpha = 0$
$RH(\alpha)_{G1A} =$	$\alpha(RH1_{\alpha_1 L} - RH1_{\alpha_0 L}) + RH1_{\alpha_0 L}$	$0 < \alpha < 1$
	$RH1_{\alpha_1 H}$	$\alpha = 1$
	$RH2_{\alpha_0 L} \wedge RH2_{\alpha_0 H}$	$\alpha = 0$
$RH(\alpha)_{G2A} =$	$\alpha(RH2_{\alpha_1 L} - RH2_{\alpha_0 L}) + RH2_{\alpha_0 L} \wedge \alpha(RH2_{\alpha_1 H} - RH2_{\alpha_0 H}) + RH2_{\alpha_0 H}$	$0 < \alpha < 1$
	$[RH2_{\alpha_1 L}, RH2_{\alpha_1 H}]$	$\alpha = 1$
	$RH3_{\alpha_0 L} \wedge RH3_{\alpha_0 H}$	$\alpha = 0$
$RH(\alpha)_{G3A} =$	$\alpha(RH3_{\alpha_1 L} - RH3_{\alpha_0 L}) + RH3_{\alpha_0 L} \wedge \alpha(RH3_{\alpha_1 H} - RH3_{G0,\max}) + RH3_{\alpha_0 H}$	$0 < \alpha < 1$
	$[RH3_{\alpha_1 L}, RH3_{\alpha_1 H}]$	$\alpha = 1$

Table 2. Membership function model for adjusting reference borders of air temperature and RH.

temperature ($\alpha_s = 0$) and optimum air temperature ($\alpha_s = 1$) are, respectively, shown in red- and green-dashed lines. In this example, the percentage of data that falls inside these three reference borders ($\alpha_s = 0$, 0.7 and 1) are 100, 92 and 41% for the early growth stage, and 100, 73, and 3% for the mature fruiting stage. These values are expressed on the plots of **Figure 9** as $Cft(T, 24, 0)_{GS1} = 1$, $Cft(T, 24, 0.7)_{GS1} = 0.92$, $Cft(T, 24, 1)_{GS1} = 0.41$, $Cft(T, 24, 0)_{GS5} = 1$, $Cft(T, 24, 0.7)_{GS5} = 0.73$, and $Cft(T, 24, 1)_{GS5} = 0.03$. In other words, $Cft(T, 24, 0.7)_{GS1} = 0.92$ and $Cft(T, 24, 0.7)_{GS5} = 0.73$ imply that for nearly 22 h (92% of the entire 24 h) of the random day at the early growth, and for 17.5 h (73% of the entire 24 h) of the random day at the mature fruiting stage, the climate controller (for this example, natural ventilation) provided the greenhouse with air temperature that was at least 70% optimal for tomato cultivation. Moreover, $Cft(T, 24, 1)_{GS1} = 0.41$ implies that at the early growth stage, the greenhouse was controlled

http://dx.doi.org/10.5772/intechopen.69972

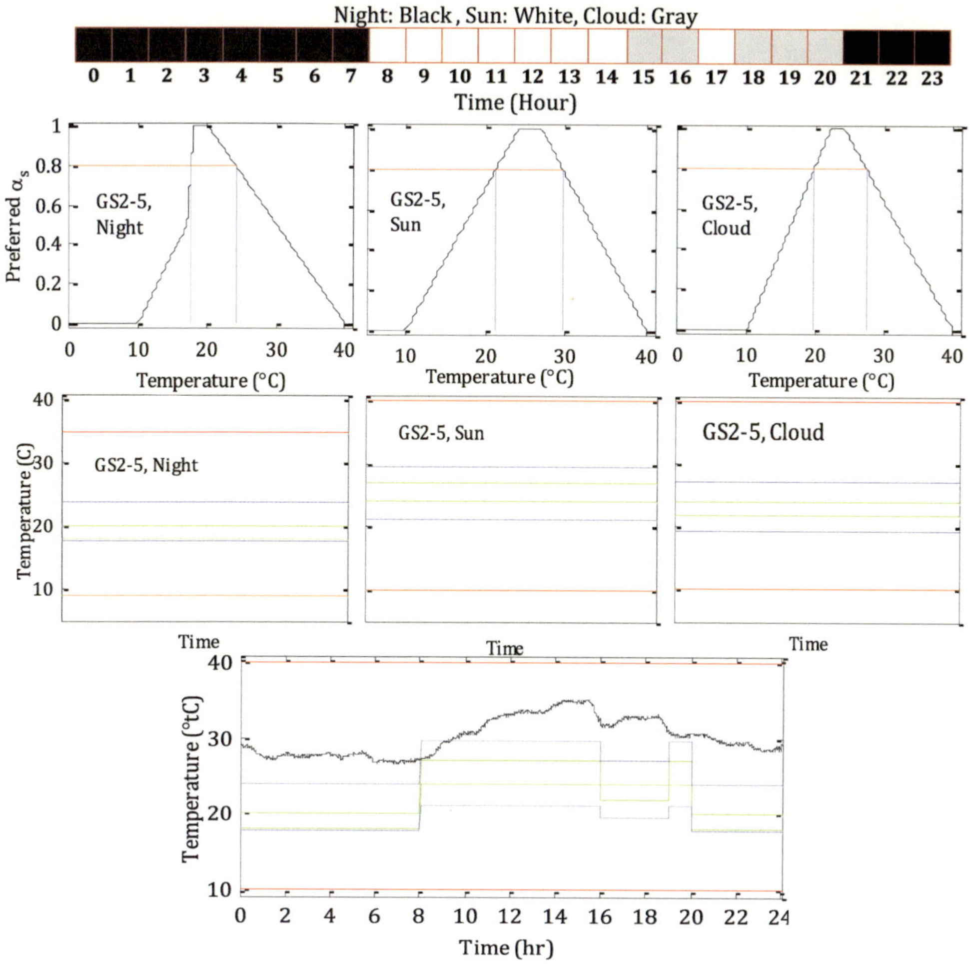

Figure 8. Demonstration of adjusting reference borders with light condition and a preferred optimality degree of $\alpha = 0.8$ for air temperature control and evaluation in a random day at the flowering to mature fruiting growth stage.

with 100% optimal air temperature for a total of 9.6 h (41% of the total 24 h, shown by green color between hours of 00:00–11:00 on the left plot of **Figure 9**). For the random day at the mature fruiting stage, it can be seen that only 3% of the air temperature response is inside $\alpha_s = 1$ reference borders (around hour 8:00 to 8:30 am).

The discussion for comfort ratio is extended to compare VPD response in three different greenhouses for a random data collection day during the flowering growth stage (GS3). The greenhouses had different covering materials and climate control system (labeled by A, B, and C in **Figure 10**, respectively, covered with net-screen mesh, polyethylene film, and polycarbonate panels). The preferred reference border for this evaluation is $\alpha_s = 0.6$ (blue-color borders). It can be observed that VPD response never crossed $\alpha = 0$ or the failure reference borders in greenhouses A and C. This can be expressed by saying that $Cft(VPD, 24, 0)_{GS3}$ was never less

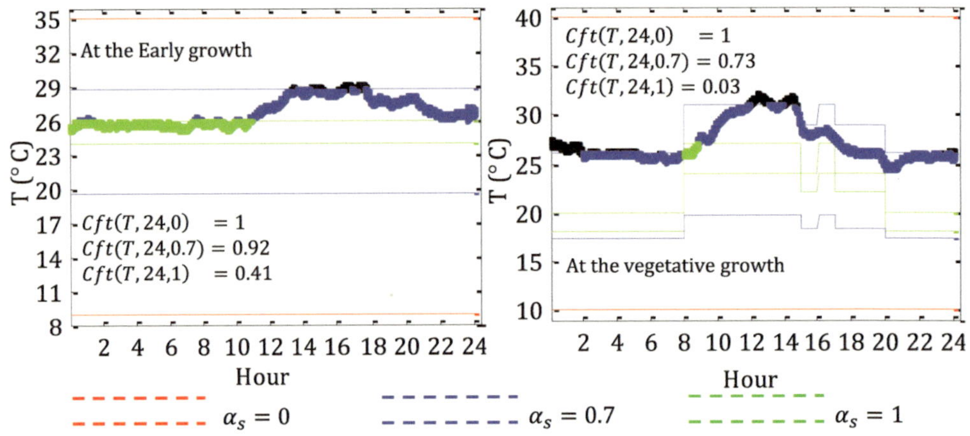

Figure 9. Demonstration of air temperature response and corresponding comfort ratios for two random days of experiment at the early growth (left) and mature fruiting stage (right) in a tropical greenhouse.

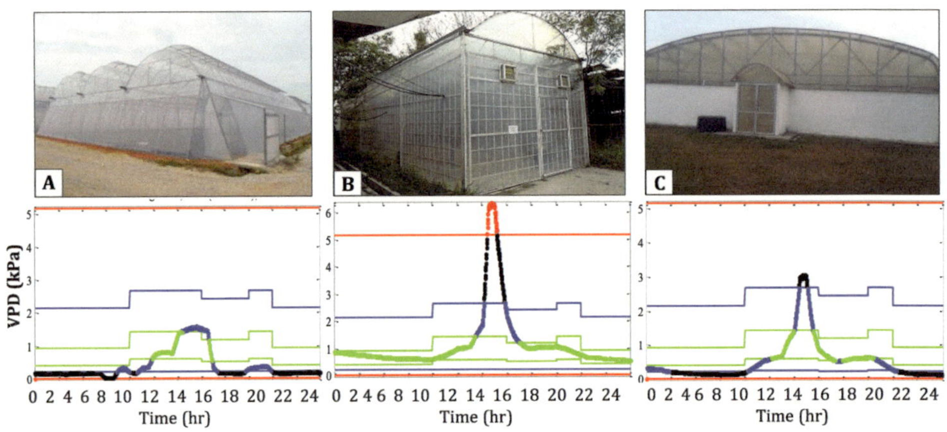

Figure 10. A comparison between comfort ratio of VPD at reference borders of $\alpha = 0$, $\alpha = 0.6$, and $\alpha = 1$ in three different greenhouses. Date of data collection March 18, 2013.

than 1 in greenhouses A and C. It should be mentioned that these two greenhouses were, respectively, operating on natural ventilation and evaporative cooling system during the experiment. According to the plots of the three greenhouses in **Figure 10**, no significant difference can be observed in their VPD responses between 0.1 and 1.2 kPa (corresponding to air temperature between 20 and 30°C, and RH between 80 and 100%); however, as air temperature starts rising above 30°C, differences in the environments start growing nonlinearly. The hourly averaged values of microclimate parameters for this experiment reveal that the major differences between these greenhouses occur between hours of 11:30 am to 4:00 pm. The mean VPD value for greenhouses B and C was equal to 2.9 and 1.19 kPa, respectively, which are less desirable for plant growth compared with the 0.97-kPa value observed from greenhouse A.

This observation indicates that as long as the outside temperature is less than 30°C, no major differences between the three greenhouses resulted. This example indicates that for this particular day of experiment, the net-screen-covered greenhouse operating on natural ventilation had a comfort ratio equal to 1 at $\alpha_s = 0.6$, which is slightly higher than $Cft(T, 24, 0.6)_{GS5} = 0.95$ of the polycarbonate panel greenhouse with evaporative cooling system. It should be noted that greenhouse C was constructed with more expensive materials, including polycarbonate panels to reduce direct sun radiation, and was operating on evaporative cooling system with large fans that consume substantial amount of electricity. This example clearly shows the potential of natural ventilation in providing more desirable response for tomato cultivation under tropical climate conditions.

3.3. Simulation of expected yield

A peer-reviewed published state-variable tomato growth model, developed by Ref. [24] in Microsoft Excel spreadsheets, was studied and implemented in MATLAB Simulink (shown by Simulink blocks in **Figures 11–13**). The objective was to provide a standalone application in a way that end users unfamiliar with programming languages and/or crop modeling would have an easier access to yield prediction in different greenhouse environments. Data from spreadsheet version of the model were used for testing the Simulink blocks and validation of the results [25]. The five state variables included in the tomato growth model of Ref. [24] were node number (N), leaf area index (LAI), total plant weight (W) or biomass, total fruit weight (W_F), and mature fruit weight (W_M). Vegetative node development is calculated on an hourly time step using greenhouse temperature (T). The state-variable equation for the rate of node development (dN/dt) is expressed by $dN/dt = N_m.f_N(T)$, where N_m is the maximum rate of node appearance per day and $f_N(T)$, is a function to reduce node development under nonoptimal temperatures on an hourly basis. Based on studies of tomato phenology, N_m was established to be 0.02083 $nodes.d^{-1}$ in the model, and the function, $f_N(T)$, is $f_N(T) = \min(1, \min(0.25 + 0.025T, 2.5 - 0.05T))$, where T is the hourly greenhouse temperature in °C. Gross hourly photosynthesis (P_h) was calculated as a function of hourly temperature, incoming solar radiation, and LAI using Eq. (1) developed by Ref. [26]. The Simulink blocks for hourly node development and hourly photosynthesis are shown in **Figure 11**. Here, D is a coefficient to convert P_h from $\mu mol(CO_2)m^{-2}.\, s^{-1}$ to $g(CH_2O)m^{-2}.\, d^{-1}$, K is the light extinction coefficient, m is the leaf light transmission coefficient, $LF_{\max}$ is the maximum leaf photosynthetic rate, $Q_e(T)$ is the leaf quantum efficiency and a function of temperature, $PPFD$ is the photosynthetic photon flux density or the level of incoming solar radiation, and $PGRED(T)$ is a function to modify P_h under suboptimal temperatures. Based on previous work with tomato growth models [24], D, K, m, and $LF_{\max}$ were set to 0.108, 0.58, 0.1, and 26, respectively. The function for $Q_e(T)$ can be expressed by $Q_e(T) = 0.084\, .\, \left(1 - 0.143 \exp\left(0.0295.(T - 23)\right)\right)$.

The function for $PGRED(T)$ was disregarded for this model because environmental conditions inside a greenhouse will not fluctuate significantly enough such that this function would have an effect on tomato growth simulations. Temperature and incoming solar radiation information necessary for computation of P_h were obtained from hourly measured data in

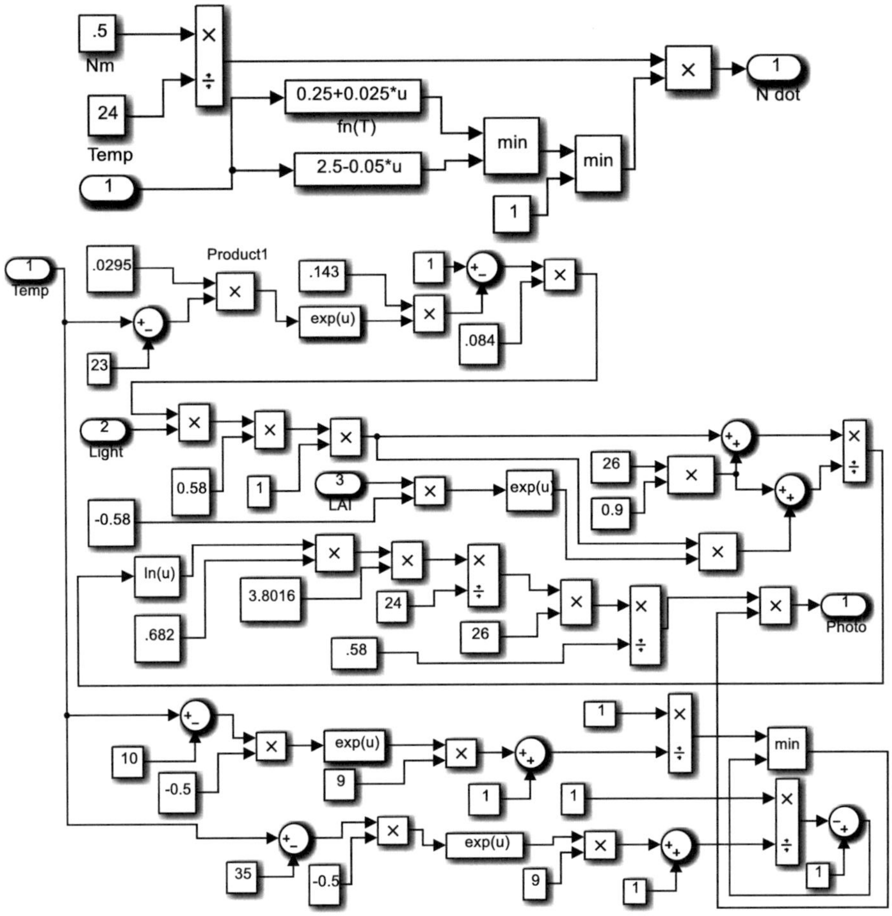

Figure 11. Simulink blocks for hourly node development and hourly photosynthesis.

the greenhouses under study, and LAI was obtained using a feedback loop in the model. Gross daily photosynthesis (P_g) was found by integrating over the 24-hourly photosynthesis calculations during each day. Hourly maintenance respiration (R_h) was computed as $R_h = r_m.Q_{10}^{(T-20)/10}$, where r_m and Q_{10} are maintenance respiration coefficients for tomato with values of 0.019 and 1.4, respectively. Daily maintenance respiration (R_m) was computed by integrating over the 24-hourly respiration calculations during the day. Vegetative node development was the only state variable computed on an hourly time step. The remaining state variables were calculated on a daily time step. The state-variable equation for computing LAI was derived from the work of [27, 28]. This state-variable equation is expressed by Eq. (2), where ρ is the plant density, $\lambda(T_d)$ is a function to reduce the rate of leaf area expansion for nonoptimal temperatures, and δ, β, and N_b are coefficients in the expolinear growth equation developed by Ref. [27]. For this work, the values for ρ, δ, β, and N_b were 3.12 plants m^{-2}, 0.038 m^{-2} $node^{-1}$, 0.169 $node^{-1}$, and

http://dx.doi.org/10.5772/intechopen.69972

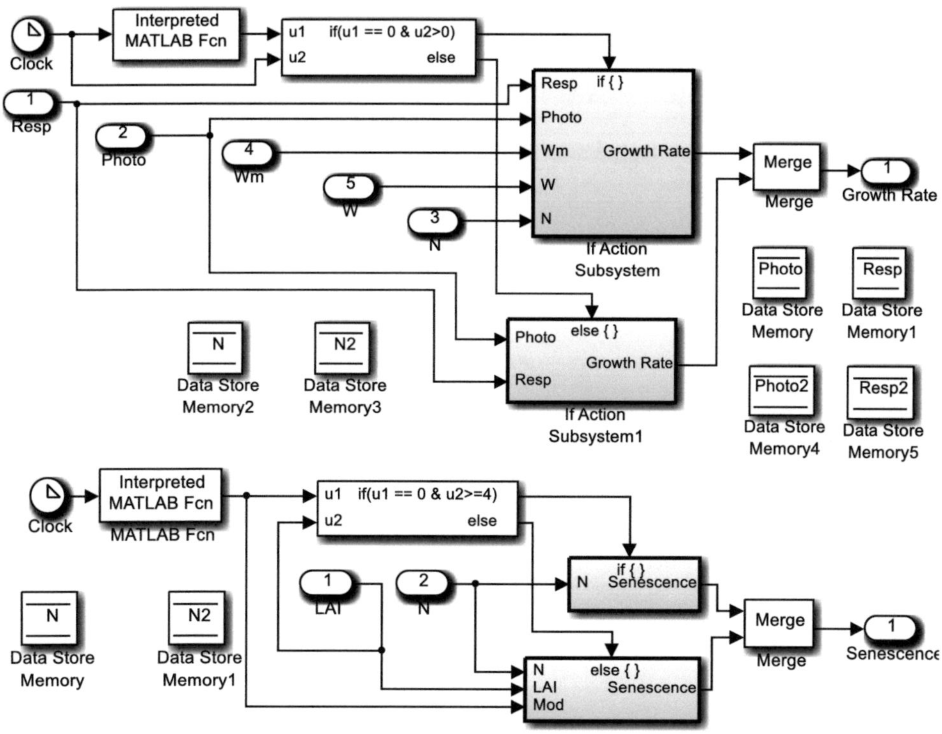

Figure 12. Simulink blocks for daily biomass accumulation and senescence.

16 nodes, respectively. The function, $\lambda(T_d)$, was not necessary for this model because temperatures within a greenhouse will not fluctuate enough for this function to significantly affect leaf area expansion simulations. The value for N is the node count at the end of the previous day, and dN/dt is the change in node count during the current day. The model assumes that when LAI reaches LAI_{max}, any additional leaf growth will be either pruned or senesced to maintain LAI at a constant value for the remainder of the growing period. For this work, the value of LAI_{max} was set to 4 as recommended by Ref. [24]. The state-variable equation for computing the accumulation of aboveground biomass (W) is based on the equation for daily plant growth (GR_{net}), that is, $GR_{net} = E.\left[P_g - R_m(W - W_M)\right].\ \left[1 - f_R(N)\right]$. Here, $(W - W_M)$ is the difference between the total aboveground biomass and the total mature fruit, and this difference represents the growing and respiring plant mass. This difference is multiplied by the daily respiration rate (R_m) to get the amount of carbon necessary for plant maintenance. Subtracting this value from the total carbon assimilated during the day (P_g) gives the total carbon available for plant growth. The coefficient, E, represents the efficiency at converting photosynthate to crop biomass, and this value was set to 0.75 in this work. The function, $f_R(N)$, determines the proportion of carbon that is partitioned to roots as a function of the number of nodes, and it can be expressed as $f_R(N) = \max(0.02, 0.18 - 0.0032.N)$. The function allows a relatively large portion of carbon to

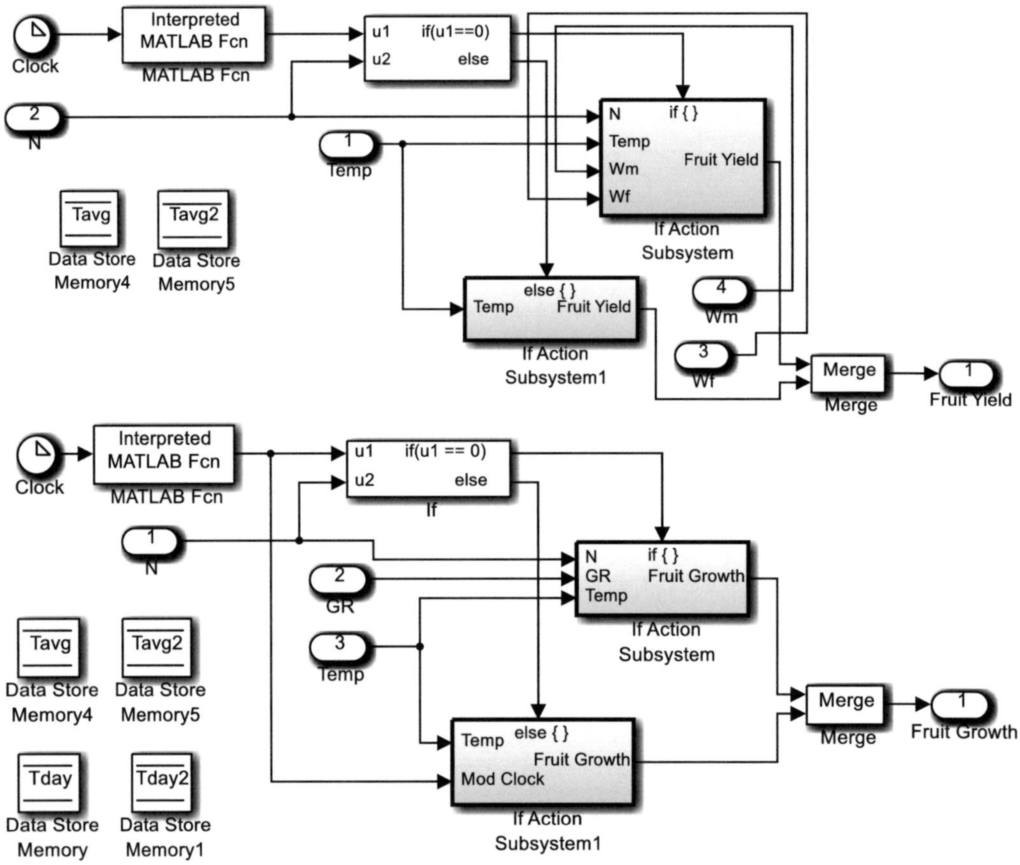

Figure 13. Simulink blocks for daily mature fruit weight and daily fruit growth.

be allocated to roots when the plant is young, and this portion tapers off to 0.02 as the plant matures. The state-variable equation for computing the accumulation of aboveground biomass (W) is $dW/dt = GR_{net} - p_1.\rho.dN/dt$, where p_1 is the dry matter weight of leaves removed per day due either to senescence or to pruning after LAI_{max} is achieved. For this work, the value of p_1 was 0 $g.node^{-1}$ before LAI_{max} was reached and 2 $g.node^{-1}$ after LAI_{max} was reached. The state-variable equation to calculate the total fruit weight (W_F) is expressed by Eq. (3). Simulink blocks for daily biomass accumulation and senescence are shown in **Figure 12**.

Here, α_F is the maximum partitioning of new growth to fruit, $f_F(T_d)$ is a function to modify partitioning to fruit according to the average daily temperature (T_d), ϑ is the transition coefficient between vegetative and full fruit growth, N_{FF} is the nodes per plant when the first fruit appears, and $g(T_{daytime})$ is a function to reduce fruit growth due to high daytime temperature. For this work, α_F, ϑ, and N_{FF} were 0.95 d^{-1}, 0.2 $node^{-1}$, and 10 nodes, respectively. The function $f_F(T_d)$ is expressed as $f_F(T_d) = \max(0, \min(1, 0.0625.(T_d - T_{min})))$, where T_{min} is the minimum temperature below which no fruit growth occurs. The function $g(T_{daytime})$ is expressed by Eq. (4) where $T_{daytime}$ is the average temperature during daylight hours and T_{crit} is the temperature above which fruit abortion begins. For tomato, T_{min} and T_{crit} are 8.5 and 24.4°C, respectively. The state-variable equation to calculate the total weight of mature fruit or the total tomato yield is expressed by Eq. (5) where $D_F(T_d)$ is a function for the rate of fruit development according to the average daily temperature, and κ_F is the development time from first fruit to first ripe fruit. For this work, κ_F was five nodes, and the function, $D_F(T_d)$, is expressed as $D_F(T_d) = 0.04 .\max(0, \min(1,\ 0.0714.(T_d - 9)))$. Mature fruit is assumed to be harvested from the plants immediately upon ripening, as shown by the subtraction of W_M during each time step from net crop growth explained by GR_{net} equation. Simulink blocks for daily mature fruit weight and daily fruit growth are shown in **Figure 13**. This description completely explicates the reduced state-variable tomato model implemented in Simulink for this project, and the state-variable equations for LAI, total biomass accumulation (dW/dt), total fruit weight (dW_F/dt), and mature fruit weight ((dW_M/dt)) are highlighted. The implemented model was validated [25] using the Lake City experiment datasets of Ref. [24] to show that the Simulink version of the model is an exact replication of the original spreadsheet version. It was then used in yield prediction from the three greenhouses shown in **Figure 10**. Results of the prediction are summarized in **Figure 14**, showing that the net-screen greenhouse operating on natural ventilation (greenhouse labeled A) had the highest yield compared with the polycarbonate panel and polyethylene film greenhouses. This result is completely consistent with results of the optimality degrees and comfort ratios obtained in the previous sections.

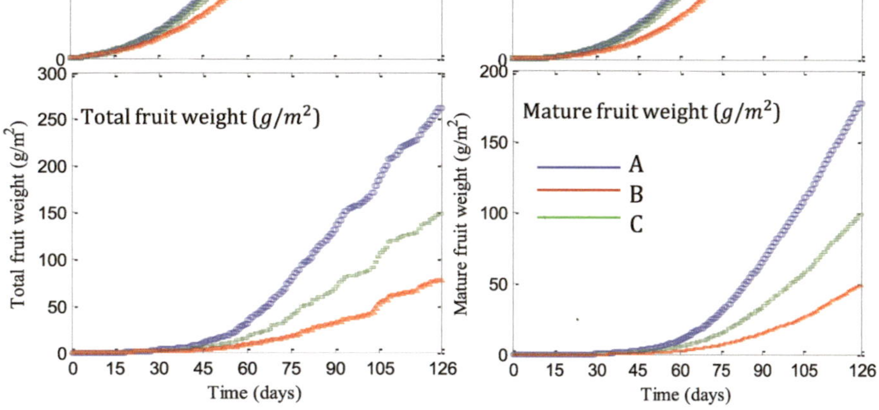

Figure 14. Simulated results with TOMGRO model for three experimental greenhouses.

4. Set-point manipulation for optimum climate control

4.1. Critical reference borders

The comfort ratio curve, denoted by Cft- curve, refers to the plot of $Cft(\mathcal{M}, t, \alpha_s)$ values calculated for all $\alpha_s = 0 : d_\alpha : 1$. It shows how much close a microclimate parameter can be controlled to different preferred reference borders. The horizontal blue-dashed line at $Cft(\mathcal{M}, t, \alpha_s) = 1$ represents 100% satisfied control objective; that is, parameter $\mathcal{M}$ is always inside reference borders of α_s. The Cft- curve can be used as a tool to demonstrate the behavior of $Cft(\mathcal{M}, t, \alpha_s)$ in different greenhouses or at different cultivation days for decision making in set-point manipulation for the climate controller. For example, it can be used in finding the largest α_s for which $Cft(M, t, \alpha_s) = 1$ (in other words, finding $\alpha_{\max}$ corresponding to the narrowest achievable reference border by the climate controller). An example is provided in **Figure 15** by plotting air temperature response for 2 consecutive days of an experiment inside a tropical greenhouse. It can be observed that the narrowest reference borders of air temperature that was completely satisfied by the climate controller in these two days are, respectively, equal to $\alpha_s = 0.55$ and $\alpha_s = 0.67$. After these points, comfort ratio starts decreasing until it arrives at its lowest value of 0.42 for both days at $\alpha_s = 1$.

Another application of the Cft- curve includes finding critical reference borders, denoted by α_{Crit} at which $\Delta = Cft(\mathcal{M}, t, \alpha_s) - Cft(\mathcal{M}, t, \alpha_s + \epsilon)$ is maximum (reference borders that cause significant loss in comfort ratio). To further explain, comfort ratios of air temperature for two distinct cases are plotted in **Figure 16**. In the first case, increasing α_s from 0.3 to 0.65 has not caused significant loss in the resulting comfort ratio. The values of $Cft(T, t, 0.3)$ and $Cft(T, t, 0.75)$ for this case are nearly the same and equal to 0.8 and 0.77. In other words, by increasing α_s from 0.3 to 0.75 to provide air temperature response that is more favored by tomato plants, performance of the controller in achieving the extra accuracy was not decreased. In a greenhouse with natural ventilation, this means that the extra 0.35 increase in α_s comes at no cost (no significant loss of response). In the case of an energy-consuming climate controller (i.e., pad-and-fan-evaporative system or swamp cooler), it means that the

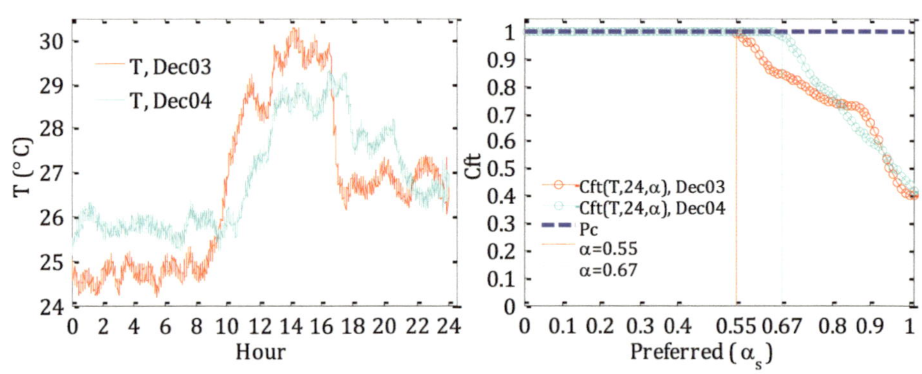

Figure 15. Comparison between air temperature responses from a tropical greenhouse in 2 days of experiment showing raw data (left), and comfort ratios (right). The controller did not satisfy 100% optimal references.

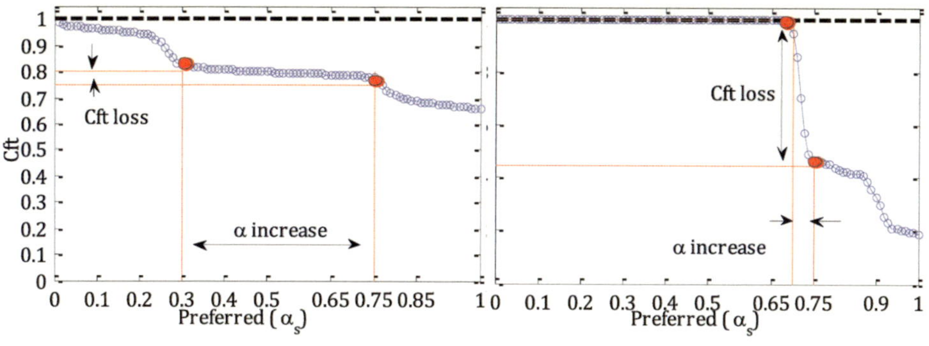

Figure 16. Comparison between comfort ratios versus α_s in 2 days of experiment in a greenhouse with evaporative cooling system for demonstration of α_{Crit}. Left: significant increase in α_s from 0.3 to 0.75 resulting in significant loss in Cft, right: slight increase in α_s from 0.7 to 0.75 causing significant loss in Cft.

cooler can be set to maintain air temperature inside a narrower reference border (by selecting $\alpha_s = 0.75$ rather than 0.3) without imposing additional energy cost. On the right plot of **Figure 16**, this situation is, however, different. Significant loss in $Cft(T, t, \alpha_s)$ can be observed for a slight increase from $\alpha_s = 0.7$ to $\alpha_s + \epsilon = 0.75$. Here, increasing α_s for as little as 0.05 has led to a sudden drop in the comfort ratio by 50% (from 1 to 0.5). The α_s at which the largest loss appear is referred to α_{Crit} and can be calculated by differentiating Cft -curve with respect to α as $\alpha_{crit} = d/d_\alpha(Cft(\mathcal{M}, t, \alpha))$.

4.2. Performance of climate controller

Plots of measured optimality degrees of a response parameter, denoted by $Opt(\mathcal{M}) = \alpha_y$, corresponding to the preferred α_s reference borders can provide a useful graphical tool to monitor performance of the climate control system. For the sake of demonstration, Cft- curves and performance curves of the climate controller for T, RH, and VPD are shown in **Figure 17**. For a perfectly control task with a preferred α_s, the control system must achieve microclimate parameter $\mathcal{M}$ that has optimality degree of at least α_s. For example, if reference borders of air temperature control are set at $\alpha_s = 0.8$, it is expected that the optimality degree of air temperature response inside the greenhouse is at least $\alpha_y = 0.8$ at any measured time. As mentioned earlier, in a 100% perfectly controlled greenhouses, the measured optimality degrees are at least equal to the preferred optimality degrees of the reference border ($\alpha_y = \alpha_s$). This is shown by the perfect control line (line of $\alpha_y = \alpha_x$) on the response plot of **Figure 17**. It should be noted that α_y can also be calculated by integrating $Cft(\mathcal{M}, t, \alpha)$ curve over $\alpha = 0$ to $\alpha = \alpha_s$ (Eq. (6)), indicating that α_y is equal to α_s only when $Cft(\mathcal{M}, t, \alpha_s)$=1. In other words, performance of a climate control system in achieving preferred reference borders of $\mathcal{M}$ is considered 100% perfect only when 100% of $\mathcal{M}$-response falls inside the α_s preferred optimal reference borders.

In controlled greenhouses, both *Cft curve* and performance curve provide a graphical assessment tool for comparing different control strategies and scenarios (i.e., microclimate responses due to different greenhouse designs, cooling systems, and covering materials at different

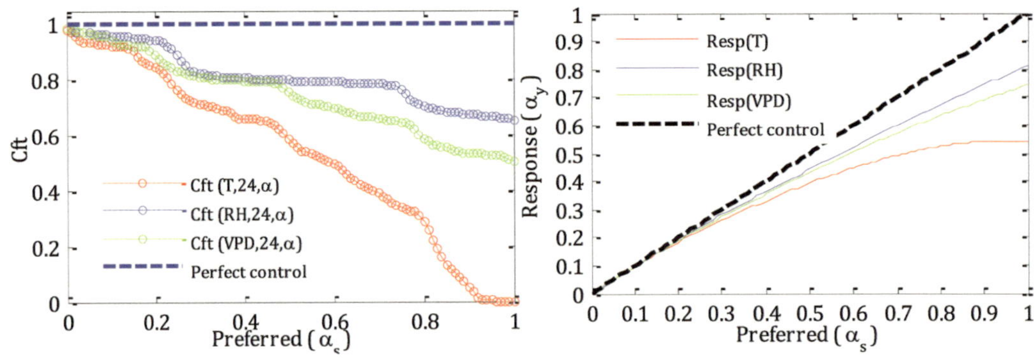

Figure 17. Comfort ratio of microclimate parameters (left) and response of the climate controller (right) at $0 \le \alpha_s \le 1$.

growth stages). The performance curve in fact reveals how much a greenhouse microclimate parameter deviates from a perfectly controlled response. Deviation of the greenhouse from this ideal line at any α_s can be used as an index factor of the perfect climate control task. The lesser deviation means the more perfect control task. Adaptability factor of the controller for microclimate parameter $\mathcal{M}$ at a preferred α_s, denoted by $ADP(\mathcal{M}, \alpha_s)$, is then defined as the ability of the controller to adapt itself with different preferred references and is calculated using Eq. (7).

4.3. Optimum reference borders

The optimum preferred reference border for parameter $\mathcal{M}$, denoted by α_{Opt}, is defined as the largest possible α_s value for which the largest $Cft(\mathcal{M}, t, \alpha)$ can be achieved. In other words, it is the value of an unknown α_i for which $Cft(\mathcal{M}, t, \alpha_i) = \beta_i$ has the minimum distance to $Cft(\mathcal{M}, t, 1) = 1$. In that sense, the cost function for this optimization problem is defined as $D_i = \sqrt{(\alpha_i - 1)^2 + (\beta_i - 1)^2}$, which is the Euclidean distant between the unknown point (α_i and β_i) on the *Cft curve* and the point of ideal microclimate ($\alpha = 1$ and $\beta = 1$). The objective is

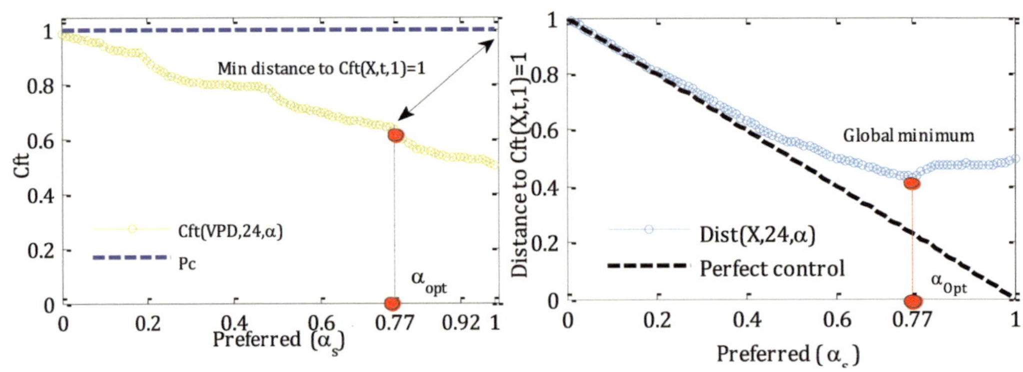

Figure 18. Demonstration of the algorithm for finding optimum preferred reference border for adjusting the climate controller. Data belongs to VPD response from a random data collection day in a tropical greenhouse experiment.

therefore to minimize this cost function by finding $0 \le \alpha_i \le 1$ value that leads to the shortest Euclidean distant (minimum $Dist_i$) to the $Cft(\mathcal{M}, t, 1) = 1$. An example is demonstrated in **Figure 18** for VPD response in a random day of experiment with $\alpha_{\text{Opt}} = 0.77$. The plot on the right side of **Figure 18** shows the values of D_i versus $0 \le \alpha_i \le 1$, and the position of α_{Opt} is shown as the global minimum point.

5. Conclusion

An adaptive management framework was designed, developed, and introduced in this research to respond to the needs for an iterative processing tool that acknowledge complexity and uncertainty in microclimate control and management. A systematic approach was presented for automatic data collection and processing with the objective to produce knowledge-based information in achieving optimum microclimate for producing high-quality and high-yield tomato. Applications of computer models were demonstrated through case-study examples for measuring and adjusting optimality degrees, comfort ratios, and prediction of the expected yield. Several applications of the framework toolboxes were demonstrated through case-study examples for evaluating and comparing microclimate parameters as well as yield prediction in different greenhouse environments. Specific applications of the optimization toolbox of the framework were discussed for evaluating and adjusting greenhouse climate controller through manipulated set points. It was shown that using adaptive greenhouse model for tropical climate condition, efficient use of natural ventilation, or shading will cause up to 70% savings on other energy-consuming cooling systems without sacrificing fruit quality or yield. The presented approach can be used in cost-benefit analysis for providing best management decisions such as site selection, optimum growing season, scheduling efficiencies, energy management with different climate control systems, and risk assessments associated with each task. Results of microclimate evaluation and yield prediction that are generated by this framework can be used in other crop models that estimate plant responses to the environment, or contribute to task-planning algorithms for hierarchical decomposition of climate management, and in economic models of tomato for energy conservation and energy efficient greenhouse crop productions. The framework can also be used as a research tool in future studies such as evaluating effects of different greenhouse designs and shapes on comfort ratios of microclimate parameters, or finding optimum combination of ventilation and evaporative cooling systems for best fruit quality and yield.

6. Technical data

The custom-designed data acquisition and control system [17] for monitoring and manipulating of the microclimate parameters was built using Arduino Uno prototype board utilizing ATmega328P (Atmel®, San Jose, CA) microcontroller on the open source Arduino Uno prototyping platform programmable in Arduino sketch environment software with C (C Compiler, Brookfield, WI), a liquid crystal display, power supply, and serial port RS-232 communication cable (bidirectional with a maximum baud speed up to 115,200 bites per seconds) for transferring and storing collected data in PC. All vital components (i.e., clock generator, 2 KB of RAM, 32 KB of flash memory for storing programs and 1 KB of EEPROM for storing parameters, a 16-MHz crystal oscillator, digital input/output pins, USB connection, power regulator, power jack, and a reset button) for operating the microcontroller, as well as direct programming and access to input/output pins, were available on the prototype board. Four arrays of HSM-20G-combined sensors modules (Shenzhen Mingjiada Electronics LTD, Futian Shenzhen, China), external micro-secure digital (SD) cardboard for storing larger amount of sensor data, output connection, sensor input, and relay circuit board for on/off control purposes were used. The data acquisition interface was tested for accuracy and reliability with available commercial models, and with a control sample data collected from airport weather station at Sultan Abdul Aziz Shah-Subang in Malaysia.

$$P_h = \frac{D\,.\,LF_{max}\,.\,PGRED(T)}{K}.ln\left[\frac{(1-m).LF_{max}+Q_e(T)\,.\,K.\,PPFD}{(1-m)\,.\,LF_{\max}+Q_e(T).\,K.\,PPFD\,.\,exp(-k\,.\,LAI)}\right] \tag{1}$$

$$\begin{cases} \dfrac{d(LAI)}{dt} = \rho.\delta.\lambda(T_d).\dfrac{exp\left[\beta.(N-N_b)\right]}{1+\exp\left[\beta.(N-N_b)\right]}.\dfrac{dN}{dt} & : LAI \leq LAI_{\max} \\ \dfrac{d(LAI)}{dt} = 0 & : LAI \geq LAI_{\max} \end{cases} \tag{2}$$

$$\frac{dW_F}{dt} = GR_{net}.\alpha_F.f_F(T_d).\left[1-\exp\left(-\vartheta(N-N_{FF})\right)\right].g\left(T_{daytime}\right)\ if\ N > N_{FF} \tag{3}$$

$$g\left(T_{daytime}\right) = \max\left(0.09, \min\left(1, 1-0.154\left(T_{daytime}-T_{crit}\right)\right)\right) \tag{4}$$

$$\frac{dW_M}{dt} = D_F(T_d).(W_F - W_M),\ if\ N > N_{FF}+\kappa_F \tag{5}$$

$$\alpha_y = \int_{\alpha=0}^{\alpha=\alpha_s} Cft(\mathcal{M}, t, \alpha).d\alpha = \sum_{i=1}^{N} Cft(\mathcal{M}, t, \alpha_i) \times \alpha_i \tag{6}$$

$$ADP(\mathcal{M}, \alpha_s) = 1-2\left(\int_{\alpha=0}^{\alpha=\alpha_s} \alpha.d\alpha - \int_{\alpha=0}^{\alpha=\alpha_s} Opt(\mathcal{M}).d\alpha\right) \tag{7}$$

Acknowledgements

The financial support provided by the University of Putra Malaysia, Grant Number GP-IPB/ 2013/9415600, and the scientific comments and suggestions from Professor Warren Dixon, Professor Jim Jones, and Professor Ray Bucklin at the University of Florida, and Professor Jan Bontsema at the Wageningen University and Research Center are duly acknowledged.

Author details

Redmond R. Shamshiri[1,2]*, Muhammad Razif Mahadi[2], Kelly R. Thorp[3], Wan Ishak Wan Ismail[2], Desa Ahmad[2] and Hasfalina Che Man[2]

*Address all correspondence to: Redmond@AdaptiveAgroTech.com

1 Department of Agricultural and Biological Engineering, University of Florida, Gainesville, FL, USA

2 Department of Biological and Agricultural Engineering, Faculty of Engineering, Universiti Putra Malaysia, Serdang, Selangor, Malaysia

3 United States Department of Agriculture, Agricultural Research Service, Maricopa, AZ, USA

References

[1] Ting, K.C. Automation for Phytomation Systems on Instrumentation and Information Technology for Bioproduction Systems. Invited lecture. Tokyo Institute of Agriculture and Technology, Tokyo, Japan. 1999.

[2] Ting KC, Fleisher DH, Rodriguez LF. Concurrent science and engineering for phytomation systems. Journal of Agricultural Meteorology. 2003;**59**(2):93–101

[3] Ting, K.C. A Systems Concept for Controlled Environment Plant Production. Resource Magazine. Volume March/April. Page 26. Published by the American Society of Agricultural and Biological Engineering (ASABE). 2013. Available online at: http://bt.e-ditionsbyfry.com/publication/index.php?i=148288&m=&l=&p=1&pre=&ver=html5#{"page":26,"issue_id":148288}Last accessed on July 10, 2017

[4] Fleisher DH, Baruh H. An optimal control strategy for crop growth in advanced life support systems. Life Support & Biosphere Science. 2001;**8**(1):43–54

[5] Fleisher DH, Ting KC. Object-oriented analysis and modeling of closed plant production systems. In: Transplant Production in the 21st Century. The Netherlands: Kluwer Academic Publishers; 2001. pp. 53–58

[6] Fleisher DH, Ting KC, Giacomelli GA. Decision support software for phytoremediation systems using rhizofiltration processes. Transactions of the Chinese Society of Agricultural Engineering. 2002;**18**(5):210–215

[7] Holling CS. Adaptive Environmental Assessment and Management. Chichester: Wiley; 1978

[8] Linkov I, Satterstrom FK, Kiker G, Batchelor C, Bridges T, Ferguson E. From comparative risk assessment to multi-criteria decision analysis and adaptive management: Recent developments and applications. Environment International. 2006;**32**(8):1072–1093

[9] Henriksen H, Barlebo H. Reflections on the use of Bayesian belief networks for adaptive management. Environmental Management. 2008;**88**(4):1025–1036

[10] Williams BK. Adaptive management of natural resources—Framework and issues. Journal of Environmental Management. 2011;**92**:1346–1353

[11] Whicker JJ, Janecky DR, Doerr TB. Adaptive management: A paradigm for remediation of public facilities following a terrorist attack. Risk Analysis. 2008;**28**(5):1445–1456

[12] Stankey GH, Bormann BT, Ryan C, Shindler B, Sturtevant V, Clark RN, Philpot C. Adaptive management and the Northwest forest plan: Rhetoric and reality. Journal of Forestry. 2003;**101**(1):40–46

[13] Stankey GH, Clark RN, Bormann BT. Adaptive management of natural resources: Theory, concepts, and management institutions. In: General Technical Report PNW-GTR-654. Portland, OR, USA: U.S. Department of Agriculture, Forest Service, Pacific Northwest Research Station; 2005

[14] Nian-Feng W, Chen J, Jie-Xian J, Li B. A conceptual framework for ecosystem management based on tradeoff analysis. Ecological Indicators 2017;**75**(April):352–361. DOI: http://dx.doi.org/10.1016/j.ecolind.2016.12.032

[15] Aiello, G., Giovino, I., Vallone, M., Catania, P., & Argento, A. A decision support system based on multisensor data fusion for sustainable greenhouse management. Journal of Cleaner Production. 2017. https://doi.org/10.1016/j.jclepro.2017.02.197

[16] McLain RJ, Lee RG. Adaptive management: Promises and pitfalls. Environmental Management. 1996;**20**(4):437–448

[17] Shamshiri R, Wan Ismail WI. Data acquisition for monitoring vapor pressure deficit in a tropical lowland shelter-house plant production. Research Journal of Applied Sciences, Engineering and Technology. Maxwell Science Publication; 2014;**7**(20)111–122

[18] Körner O, Straten GV. Decision support for dynamic greenhouse climate control strategies. Computers and Electronics in Agriculture. 2008;**60**:18–30

[19] Berenguel M, Yebra L J, Rodríguez F. Adaptive control strategies for greenhouse temperature control. European Control Conference (ECC) IEEE. 2003. pp. 2747–2752

[20] El-Attal, A. H. Decision model for hydroponic tomato production (HYTODMOD) using utility theory (Doctoral dissertation, The Ohio State University). 1995.

[21] Short TH, Draper CM, Donnell MA. Web-based decision support system for hydroponic vegetable production. Acta Horticulturae (ISHS). 2005;**691**:867–870

[22] Shamshiri, R. Measuring optimality degrees of microclimate parameters in protected cultivation of tomato under tropical climate condition. Measurement. 2017;**106**:236–244

[23] Shamshiri, R., Che Man, H., Zakaria, A.J., Beveren, P.V., Wan Ismail, W.I. and Ahmad, D. Membership function model for defining optimality of vapor pressure deficit in closed-field cultivation of tomato. Acta Hortic. 2017:**1152**:281–290 DOI: 10.17660/ActaHortic.2017.1152.38 https://doi.org/10.17660/ActaHortic.2017.1152.38

[24] Jones, J. W., Kenig, A., & Vallejos, C. E. Reduced State–Variable Tomato Growth Model. Transactions of the ASAE. 1999:**42**(1):255–265 Doi: 10.13031/2013.13203

[25] Shamshiri R, Ahmad DB, Wan Ismail WI, Hasfalina CM, Zakaria AJ, Yamin M. Evaluation of the reduced state-variable TOMGRO model using boundary data. In: 2016 ASABE International Meeting, no. 2454205. 2016. pp. 2–12. DOI: 10.13031/aim.20162454205

[26] Acock B, Charles-Edwards DA, Fitter DJ, Hand DW, Ludwig LJ, Wilson JW, Withers AC. The contribution of leaves from different levels within a tomato crop to canopy net photosynthesis: An experimental examination of two canopy models. Journal of Experimental Botany. 1978;**29**(111):815–827

[27] Goudriaan J, Monteith JL. Mathematical function for crop growth based on light interception and leaf area expansion. Annals of Botany. 1990;**66**:695–701

[28] Vallejos, C. E., Jones J. W., and Williams F. W. High temperature tomato experiments, Ch. II-3. In Optimal Environmental Control for Indeterminate Greenhouse Crops, eds. Seginer I., Jones J.W., Gutman P. and Vallejos C. E. BARD Research Report No. IS-1995-91RC. Haifa, Israel: Agricultural Engineering Dept., Technion. 1997.